WiMAX: Technology and Applications

WiMAX: Technology and Applications

Edited by
Tristan Powell

Larsen & Keller
www.larsen-keller.com

WiMAX: Technology and Applications
Edited by Tristan Powell
ISBN: 978-1-63549-297-2 (Hardback)

© 2017 Larsen & Keller

 Larsen & Keller

Published by Larsen and Keller Education,
5 Penn Plaza,
19th Floor,
New York, NY 10001, USA

Cataloging-in-Publication Data

WiMAX : technology and applications / edited by Tristan Powell.
    p. cm.
Includes bibliographical references and index.
ISBN 978-1-63549-297-2
1. IEEE 802.16 (Standard). 2. Wireless metropolitan area networks. 3. Wireless communication systems.
4. Broadband communication systems. I. Powell, Tristan.
TK5103.2 .W56 2017
621.384--dc23

The publisher's policy is to use permanent paper from mills that operate a sustainable forestry policy. Furthermore, the publisher ensures that the text paper and cover boards used have met acceptable environmental accreditation standards.

Printed and bound in the United States of America.

For more information regarding Larsen and Keller Education and its products, please visit the publisher's website www.larsen-keller.com

Table of Contents

Preface

Worldwide Interoperability for Microwave Access or WiMAX are communication standards for wireless communications. It is widely seen as a suitable upgrade to existing technologies that are being implemented to transmit data and information. It is used for Internet access, wireless backhaul technology and triple playing. This book provides comprehensive insights into the field of WiMAX. It is compiled in such a manner, that it provides in-depth knowledge about the theory and practice of this technology. Some of the diverse topics covered in this textbook address the varied branches that fall under this category. Different approaches, evaluations and methodologies and advanced studies on WiMAX technology have been included in it. As this field is emerging at a rapid pace, the contents of this text will help the readers understand the modern concepts and applications of the subject. Those in search of information to further their knowledge will be greatly assisted by it.

To facilitate a deeper understanding of the contents of this book a short introduction of every chapter is written below:

Chapter 1- Worldwide Interoperability for Microwave Access or WiMAX is a wireless technology based on wireless MAN that, when compared to Wi-Fi, is able to provide higher speeds over larger distances to a greater number of users. The possibilities of WiMAX are endless and companies like Motorola, Sprint Nextel already started exploring the technology as early as 2008. This introduction explores WiMAX with the intention of providing the reader with an overview of this cutting-edge wireless communication.

Chapter 2- Gone are the days of wired devices that were cumbersome and lacked mobility. The common wireless technologies in use nowadays are Bluetooth, Wi-Fi and mobile phones. This chapter discusses wireless technology, wireless network and wireless mesh network. A section is committed to examining the comparison of wireless data standards. The topics discussed in the chapter are of great importance to broaden the existing knowledge on wireless technology.

Chapter 3- To provide a better understanding of WiMAX technology, it is of utmost importance that key concepts like Yota Egg, Digital subscriber line, IEEE 802.16 and WiMAX MIMO be understood by the reader. These devices provide WiMAX capability in various ways. This chapter discusses these key concepts with technical specifications, discerning characteristics and definitions.

Chapter 4- WiMAX is faster, efficient, serves remote areas and can support more users than Wi-Fi and this is the reason it is heralded as the technology of the future. Exploiting this potential, services like Internet access, IPTV and triple play have incorporated WiMAX technology and are thus able to provide cheaper, faster and more cost effective transmission at low spectrums. This chapter reviews these services with the objective of helping the reader understand the limitless potentiality of WiMAX technology.

Chapter 5- WiMAX has found use in wireless backhaul technology, hotspot and wireless LAN networks. This chapter studies these applications of WiMAX and makes the reader familiar with how WiMAX has rendered bigger prospects to these technologies and networks. This chapter discusses the methods of WiMAX in a critical manner providing key analysis to the subject matter.

Chapter 6- Several new services have emerged that utilize WiMAX like WiBro, cognitive radio, category 5 cable, high speed packet access, last mile, wireless local loop, customer-premises equipment etc. The chapter discusses the difference that WiMAX has brought to these services and how these services are revolutionizing the transmission of wireless communication. The aspects elucidated in this chapter are of vital importance, and provide a better understanding of WiMAX technology.

Chapter 7- There are several wireless technologies that are contending with WiMAX some of which are Local Multipoint Distribution Service, Wireless broadband, LTE (telecommunication), CDMA2000, UMTS (telecommunication) etc. WiMAX has distinct advantages over the other competing technologies and this chapter is a comparative study of these advantages.

Chapter 8- VoIP technology provides voice communications over the Internet and other Internet Protocol networks. These technologies enhance WiMAX technologies. This chapter is dedicated to the examination of Media Gateway Control Protocol, Simple Gateway Control Protocol, IEEE 802.21, IEEE 802.11u, Session Initiation Protocol, Mobile VoIP etc. This section provides the reader with a comprehensive study of technologies using VoIP.

Chapter 9- Bit rates are the indicators of speed and the higher the bit rate the faster the network. It is essential to study the bit rates of devices to understand the most efficient technology available and also as a comparison tool. The content of this chapter is designed to facilitate a better understanding of the bit rates of various devices that are currently in use. The topics discussed in the chapter are of great importance to broaden the existing knowledge on WiMAX technology.

Finally, I would like to thank the entire team involved since the inception of this book for their valuable time and contribution. This book would not have been possible without their efforts. I would also like to thank my friends and family for their constant support.

Editor

Introduction to WiMAX Technology

Worldwide Interoperability for Microwave Access or WiMAX is a wireless technology based on wireless MAN that, when compared to Wi-Fi, is able to provide higher speeds over larger distances to a greater number of users. The possibilities of WiMAX are endless and companies like Motorola, Sprint Nextel already started exploring the technology as early as 2008. This introduction explores WiMAX with the intention of providing the reader with an overview of this cutting-edge wireless communication.

WiMAX (Worldwide Interoperability for Microwave Access) is a family of wireless communication standards based on the IEEE 802.16 set of standards, which provide multiple physical layer (PHY) and Media Access Control (MAC) options.

WiMAX base station equipment with a sector antenna and wireless modem on top

The name "WiMAX" was created by the WiMAX Forum, which was formed in June 2001 to promote conformity and interoperability of the standard, including the definition of predefined system profiles for commercial vendors. The forum describes WiMAX as "a standards-based technology enabling the delivery of last mile wireless broadband access as an alternative to cable and DSL". IEEE 802.16m or WirelessMAN-Advanced is a candidate for the 4G, in competition with the LTE Advanced standard.

WiMAX was initially designed to provide 30 to 40 megabit-per-second data rates, with the 2011 update providing up to 1 Gbit/s for fixed stations.

Terminology

WiMAX refers to interoperable implementations of the IEEE 802.16 family of wireless-networks standards ratified by the WiMAX Forum. (Similarly, Wi-Fi refers to interoperable implementations of the IEEE 802.11 Wireless LAN standards certified by the Wi-Fi Alliance.) WiMAX Forum certification allows vendors to sell fixed or mobile products as WiMAX certified, thus ensuring a level of interoperability with other certified products, as long as they fit the same profile.

The original IEEE 802.16 standard (now called "Fixed WiMAX") was published in 2001. WiMAX adopted some of its technology from WiBro, a service marketed in Korea.

Mobile WiMAX (originally based on 802.16e-2005) is the revision that was deployed in many countries, and is the basis for future revisions such as 802.16m-2011.

WiMAX is sometimes referred to as "Wi-Fi on steroids" and can be used for a number of applications including broadband connections, cellular backhaul, hotspots, etc. It is similar to Wi-Fi, but it can enable usage at much greater distances.

Uses of WiMAX

The bandwidth and range of WiMAX make it suitable for the following potential applications:

- Providing portable mobile broadband connectivity across cities and countries through a variety of devices.

- Providing a wireless alternative to cable and digital subscriber line (DSL) for "last mile" broadband access.

- Providing data, telecommunications (VoIP) and IPTV services (triple play).

- Providing a source of Internet connectivity as part of a business continuity plan.

- Smart grids and metering

Internet Access

WiMAX can provide at-home or mobile Internet access across whole cities or countries. In many cases this has resulted in competition in markets which typically only had access through an existing incumbent DSL (or similar) operator.

Additionally, given the relatively low costs associated with the deployment of a WiMAX network (in comparison with 3G, HSDPA, xDSL, HFC or FTTx), it is now economically viable to provide last-mile broadband Internet access in remote locations.

Middle-mile Backhaul to Fiber Networks

Mobile WiMAX was a replacement candidate for cellular phone technologies such as GSM and CDMA, or can be used as an overlay to increase capacity. Fixed WiMAX is also considered as a wireless backhaul technology for 2G, 3G, and 4G networks in both developed and developing nations.

In North America, backhaul for urban operations is typically provided via one or more copper wire line connections, whereas remote cellular operations are sometimes backhauled via satellite. In other regions, urban and rural backhaul is usually provided by microwave links. (The exception to this is where the network is operated by an incumbent with ready access to the copper network.) WiMAX has more substantial backhaul bandwidth requirements than legacy cellular applications. Consequently, the use of wireless microwave backhaul is on the rise in North America and existing microwave backhaul links in all regions are being upgraded. Capacities of between 34 Mbit/s and 1 Gbit/s are routinely being deployed with latencies in the order of 1 ms.

In many cases, operators are aggregating sites using wireless technology and then presenting traffic on to fiber networks where convenient. WiMAX in this application competes with microwave radio, E-line and simple extension of the fiber network itself.

Triple-play

WiMAX directly supports the technologies that make triple-play service offerings possible (such as Quality of Service and Multicasting). These are inherent to the WiMAX standard rather than being added on as Carrier Ethernet is to Ethernet.

On May 7, 2008 in the United States, Sprint Nextel, Google, Intel, Comcast, Bright House, and Time Warner announced a pooling of an average of 120 MHz of spectrum and merged with Clearwire to market the service. The new company hopes to benefit from combined services offerings and network resources as a springboard past its competitors. The cable companies will provide media services to other partners while gaining access to the wireless network as a Mobile virtual network operator to provide triple-play services.

Some analysts questioned how the deal will work out: Although fixed-mobile convergence has been a recognized factor in the industry, prior attempts to form partnerships among wireless and cable companies have generally failed to lead to significant benefits to the participants. Other analysts point out that as wireless progresses to higher bandwidth, it inevitably competes more directly with cable and DSL, inspiring competitors into collaboration. Also, as wireless broadband networks grow denser and usage habits shift, the need for increased backhaul and media service will accelerate, therefore the opportunity to leverage cable assets is expected to increase.

Connecting

Devices that provide connectivity to a WiMAX network are known as subscriber stations (SS).

Portable units include handsets (similar to cellular smartphones); PC peripherals (PC Cards or USB dongles); and embedded devices in laptops, which are now available for Wi-Fi services. In addition, there is much emphasis by operators on consumer electronics devices such as Gaming consoles, MP3 players and similar devices. WiMAX is more similar to Wi-Fi than to other 3G cellular technologies.

A WiMAX USB modem for mobile access to the Internet

The WiMAX Forum website provides a list of certified devices. However, this is not a complete list of devices available as certified modules are embedded into laptops, MIDs (Mobile Internet devices), and other private labeled devices.

Gateways

WiMAX gateway devices are available as both indoor and outdoor versions from several manufacturers including Vecima Networks, Alvarion, Airspan, ZyXEL, Huawei, and Motorola. *The list of deployed WiMAX networks and WiMAX Forum membership list provide more links to specific vendors, products and installations. The list of vendors and networks is not comprehensive and is not intended as an endorsement of these companies above others.*

Many of the WiMAX gateways that are offered by manufactures such as these are standalone self-install indoor units. Such devices typically sit near the customer's window with the best signal, and provide:

- An integrated Wi-Fi access point to provide the WiMAX Internet connectivity to multiple devices throughout the home or business.

- Ethernet ports to connect directly to a computer, router, printer or DVR on a local wired network.

- One or two analog telephone jacks to connect a land-line phone and take advantage of VoIP.

Indoor gateways are convenient, but radio losses mean that the subscriber may need to be significantly closer to the WiMAX base station than with professionally installed external units.

Outdoor units are roughly the size of a laptop PC, and their installation is comparable to the installation of a residential satellite dish. A higher-gain directional outdoor unit will generally result in greatly increased range and throughput but with the obvious loss of practical mobility of the unit.

External Modems

USB can provide connectivity to a WiMAX network through what is called a dongle.

Generally these devices are connected to a notebook or net book computer. Dongles typically have omnidirectional antennas which are of lower gain compared to other devices. As such these devices are best used in areas of good coverage.

Mobile Phones

HTC announced the first WiMAX enabled mobile phone, the Max 4G, on November 12, 2008. The device was only available to certain markets in Russia on the Yota network.

HTC and Sprint Nextel released the second WiMAX enabled mobile phone, the EVO 4G, March 23, 2010 at the CTIA conference in Las Vegas. The device, made available on June 4, 2010, is capable of both EV-DO(3G) and WiMAX(pre-4G) as well as simultaneous data & voice sessions. Sprint Nextel announced at CES 2012 that it will no longer be offering devices using the WiMAX technology due to financial circumstances, instead, along with its network partner Clearwire, Sprint Nextel will roll out a 4G network deciding to shift and utilize LTE 4G technology instead.

Technical Information

The IEEE 802.16 Standard

WiMAX is based upon IEEE Std 802.16e-2005, approved in December 2005. It is a supplement to the IEEE Std 802.16-2004, and so the actual standard is 802.16-2004 as amended by 802.16e-2005. Thus, these specifications need to be considered together.

IEEE 802.16e-2005 improves upon IEEE 802.16-2004 by:

- Adding support for mobility (soft and hard handover between base stations). This is seen as one of the most important aspects of 802.16e-2005, and is the very basis of Mobile WiMAX.

- Scaling of the fast Fourier transform (FFT) to the channel bandwidth in order to keep the carrier spacing constant across different channel bandwidths (typically 1.25 MHz, 5 MHz, 10 MHz or 20 MHz). Constant carrier spacing results in a higher spectrum efficiency in wide channels, and a cost reduction in narrow channels. Also known as scalable OFDMA (SOFDMA). Other bands not multiples of 1.25 MHz are defined in the standard, but because the allowed FFT subcarrier numbers are only 128, 512, 1024 and 2048, other frequency bands will not have exactly the same carrier spacing, which might not be optimal for implementations. Carrier spacing is 10.94 kHz.

- Advanced antenna diversity schemes, and hybrid automatic repeat-request (HARQ)

- Adaptive antenna systems (AAS) and MIMO technology

- Denser sub-channelization, thereby improving indoor penetration

- Intro and low-density parity check (LDPC)

- Introducing downlink sub-channelization, allowing administrators to trade coverage for capacity or vice versa

- Adding an extra quality of service (QoS) class for VoIP applications.

SOFDMA (used in 802.16e-2005) and OFDM256 (802.16d) are not compatible thus equipment will have to be replaced if an operator is to move to the later standard (e.g., Fixed WiMAX to Mobile WiMAX).

Physical Layer

The original version of the standard on which WiMAX is based (IEEE 802.16) specified a physical layer operating in the 10 to 66 GHz range. 802.16a, updated in 2004 to 802.16-2004, added specifications for the 2 to 11 GHz range. 802.16-2004 was updated by 802.16e-2005 in 2005 and uses scalable orthogonal frequency division multiple access (SOFDMA), as opposed to the fixed orthogonal frequency division multiplexing (OFDM) version with 256 sub-carriers (of which 200 are used) in 802.16d. More advanced versions, including 802.16e, also bring multiple an-tenna support through MIMO. (WiMAX MIMO) This brings potential benefits in terms of coverage, self installation, power consumption, frequency reuse and bandwidth efficiency. WiMax is the most energy-efficient pre-4G technique among LTE and HSPA+.

Media Access Control Layer

The WiMAX MAC uses a scheduling algorithm for which the subscriber station needs to compete only once for initial entry into the network. After network entry is allowed, the subscriber station is allocated an access slot by the base station. The time slot can enlarge and contract, but remains assigned to the subscriber station, which means that other subscribers cannot use it. In addition to being stable under overload and over-subscription, the scheduling algorithm can also be more bandwidth efficient. The scheduling algorithm also allows the base station to control Quality of Service (QoS) parameters by balancing the time-slot assignments among the application needs of the subscriber station.

Specifications

As a standard intended to satisfy needs of next-generation data networks (4G), WiMAX is distinguished by its dynamic burst algorithm modulation adaptive to the physical environment the RF signal travels through. Modulation is chosen to be more spectrally efficient (more bits per OFDM/SOFDMA symbol). That is, when the bursts have a high signal strength and a high carrier to noise plus interference ratio (CINR), they can be more easily decoded using digital signal processing (DSP). In contrast, operating in less favorable environments for RF communication, the system automatically steps down to a more robust mode (burst profile) which means fewer bits per OFDM/SOFDMA symbol; with the advantage that power per bit is higher and therefore simpler accurate signal processing can be performed.

Burst profiles are used inverse (algorithmically dynamic) to low signal attenuation; meaning throughput between clients and the base station is determined largely by distance. Maximum distance is achieved by the use of the most robust burst setting; that is, the profile with the largest MAC frame allocation trade-off requiring more symbols (a larger portion of the MAC frame) to be allocated in transmitting a given amount of data than if the client were closer to the base station.

The client's MAC frame and their individual burst profiles are defined as well as the specific time allocation. However, even if this is done automatically then the practical deployment should avoid high interference and multipath environments. The reason for which is obviously that too much interference causes the network to function poorly and can also misrepresent the capability of the network.

The system is complex to deploy as it is necessary to track not only the signal strength and CINR (as in systems like GSM) but also how the available frequencies will be dynamically assigned (resulting in dynamic changes to the available bandwidth.) This could lead to cluttered frequencies with slow response times or lost frames.

As a result, the system has to be initially designed in consensus with the base station product team to accurately project frequency use, interference, and general product functionality.

The Asia-Pacific region has surpassed the North American region in terms of 4G broadband wireless subscribers. There were around 1.7 million pre-WiMAX and WiMAX customers in Asia - 29% of the overall market - compared to 1.4 million in the USA and Canada.

Integration with an IP-based Network

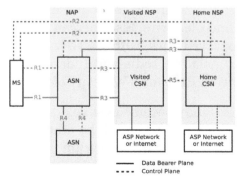

The WiMAX Forum architecture

The WiMAX Forum has proposed an architecture that defines how a WiMAX network can be connected with an IP based core network, which is typically chosen by operators that serve as Internet Service Providers (ISP); Nevertheless, the WiMAX BS provide seamless integration capabilities with other types of architectures as with packet switched Mobile Networks.

The WiMAX forum proposal defines a number of components, plus some of the interconnections (or reference points) between these, labeled R1 to R5 and R8:

- SS/MS: the Subscriber Station/Mobile Station

- ASN: the Access Service Network

- BS: Base station, part of the ASN

- ASN-GW: the ASN Gateway, part of the ASN

- CSN: the Connectivity Service Network

- HA: Home Agent, part of the CSN

- AAA: Authentication, Authorization and Accounting Server, part of the CSN

- NAP: a Network Access Provider

- NSP: a Network Service Provider

It is important to note that the functional architecture can be designed into various hardware configurations rather than fixed configurations. For example, the architecture is flexible enough to allow remote/mobile stations of varying scale and functionality and Base Stations of varying size - e.g. femto, pico, and mini BS as well as macros.

Spectrum Allocation

There is no uniform global licensed spectrum for WiMAX, however the WiMAX Forum has published three licensed spectrum profiles: 2.3 GHz, 2.5 GHz and 3.5 GHz, in an effort to drive standardisation and decrease cost.

In the USA, the biggest segment available is around 2.5 GHz, and is already assigned, primarily to Sprint Nextel and Clearwire. Elsewhere in the world, the most-likely bands used will be the Forum approved ones, with 2.3 GHz probably being most important in Asia. Some countries in Asia like India and Indonesia will use a mix of 2.5 GHz, 3.3 GHz and other frequencies. Pakistan's Wateen Telecom uses 3.5 GHz.

Analog TV bands (700 MHz) may become available for WiMAX usage, but await the complete roll out of digital TV, and there will be other uses suggested for that spectrum. In the USA the FCC auction for this spectrum began in January 2008 and, as a result, the biggest share of the spectrum went to Verizon Wireless and the next biggest to AT&T. Both of these companies have stated their intention of supporting LTE, a technology which competes directly with WiMAX. EU commissioner Viviane Reding has suggested re-allocation of 500–800 MHz spectrum for wireless communication, including WiMAX.

WiMAX profiles define channel size, TDD/FDD and other necessary attributes in order to have inter-operating products. The current fixed profiles are defined for both TDD and FDD profiles. At this point, all of the mobile profiles are TDD only. The fixed profiles have channel sizes of 3.5 MHz, 5 MHz, 7 MHz and 10 MHz. The mobile profiles are 5 MHz, 8.75 MHz and 10 MHz. (Note: the 802.16 standard allows a far wider variety of channels, but only the above subsets are supported as WiMAX profiles.)

Since October 2007, the Radio communication Sector of the International Telecommunication Union (ITU-R) has decided to include WiMAX technology in the IMT-2000 set of standards. This enables spectrum owners (specifically in the 2.5-2.69 GHz band at this stage) to use WiMAX equipment in any country that recognizes the IMT-2000.

Spectral Efficiency and Advantages

One of the significant advantages of advanced wireless systems such as WiMAX is spectral efficiency. For example, 802.16-2004 (fixed) has a spectral efficiency of 3.7 (bit/s)/Hertz, and other 3.5–4G wireless systems offer spectral efficiencies that are similar to within a few tenths of a percent. The notable advantage of WiMAX comes from combining SOFDMA with smart antenna technologies. This multiplies the effective spectral efficiency through multiple reuse and smart network deployment topologies. The direct use of frequency domain organization simplifies designs using MIMO-AAS compared to CDMA/WCDMA methods, resulting in more effective systems.

Another advantages of WiMAX, is a relatively new technology that enables communication over a maximum distance of 30 miles – compared to 300 feet for WiFi. Of course, the longer the distance, the slower the speed, but it's still faster and has a longer range than WiFi. Ideally, speeds of around 10MBps could be achieved with a range of 1 – 6 miles.

The reason why some telecommunication providers are quite excited about the prospects for WiMAX is that mobile users could use it as a faster and longer range alternative to WiFi and corporate or home users could use it in a fixed environment as a replacement or backup to DSL.

Companies will begin to use WiMAX to communicate from office to office, relatively near to each other and provide campus wide wireless connectivity to employees. Employee's computers will need to use new WiMAX cards to connect to these new networks. Next, or at the same time, public places such as airports, parks and coffee shops will be outfitted with WiMAX access points. WiMAX has been very successful as it's easy to use, low cost, and relatively fast.

While WiMAX has its benefits, as people download more and larger files, upload more data (such as voice calls, images and videos) and have longer distance needs – the limits of WiFi are apparent.

Inherent limitations

WiMAX cannot deliver 70 Mbit/s over 50 km (31 mi). Like all wireless technologies, WiMAX can operate at higher bitrates or over longer distances but not both. Operating at the maximum range of 50 km (31 mi) increases bit error rate and thus results in a much lower bitrate. Conversely, reducing the range (to under 1 km) allows a device to operate at higher bitrates.

A city-wide deployment of WiMAX in Perth, Australia demonstrated that customers at the cell-edge with an indoor Customer-premises equipment (CPE) typically obtain speeds of around 1–4 Mbit/s, with users closer to the cell site obtaining speeds of up to 30 Mbit/s. Like all wireless systems, available bandwidth is shared between users in a given radio sector, so performance could deteriorate in the case of many active users in a single sector. However, with adequate capacity planning and the use of WiMAX's Quality of Service, a minimum guaranteed throughput for each subscriber can be put in place. In practice, most users will have a range of 4-8 Mbit/s services and additional radio cards will be added to the base station to increase the number of users that may be served as required.

Silicon Implementations

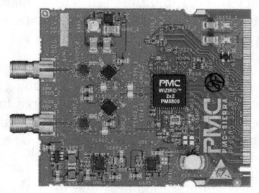

Picture of a WiMAX MIMO board

A number of specialized companies produced baseband ICs and integrated RFICs for WiMAX Subscriber Stations in the 2.3, 2.5 and 3.5 GHz bands (refer to 'Spectrum allocation' above). These companies include, but are not limited to, Beceem, Sequans, and PicoChip.

Comparison

Comparisons and confusion between WiMAX and Wi-Fi are frequent, because both are related to wireless connectivity and Internet access.

- WiMAX is a long range system, covering many kilometres, that uses licensed or unlicensed spectrum to deliver connection to a network, in most cases the Internet.

- Wi-Fi uses the 2.4 GHz, 3 GHz, 5 GHz, and 60 GHz radio frequency bands to provide access to a local network.

- Wi-Fi is more popular in end-user devices.

- Wi-Fi runs on the Media Access Control's CSMA/CA protocol, which is connectionless and contention based, whereas WiMAX runs a connection-oriented MAC.

- WiMAX and Wi-Fi have quite different quality of service (QoS) mechanisms:

  o WiMAX uses a QoS mechanism based on connections between the base station and the user device. Each connection is based on specific scheduling algorithms.

  o Wi-Fi uses contention access — all subscriber stations that wish to pass data through a wireless access point (AP) are competing for the AP's attention on a random interrupt basis. This can cause subscriber stations distant from the AP to be repeatedly interrupted by closer stations, greatly reducing their throughput.

- Both IEEE 802.11, which includes Wi-Fi, and IEEE 802.16, which includes Wi-MAX, define Peer-to-Peer (P2P) and wireless ad hoc networks, where an end user communicates to users or servers on another Local Area Network (LAN) using its access point or base station. However, 802.11 supports also direct ad hoc or peer to peer networking between end user devices without an access point while 802.16 end user devices must be in range of the base station.

Although Wi-Fi and WiMAX are designed for different situations, they are complementary. WiMAX network operators typically provide a WiMAX Subscriber Unit that connects to the metropolitan WiMAX network and provides Wi-Fi connectivity within the home or business for local devices, *e.g.*, computers, Wi-Fi handsets and smartphones. This enables the user to place the WiMAX Subscriber Unit in the best reception area, such as a window, and still be able to use the WiMAX network from any place within their residence.

The local area network inside one's house or business would operate as with any other wired or wireless network. If one were to connect the WiMAX Subscriber Unit directly to a WiMAX-enabled computer, that would limit access to a single device. As an alternative for a LAN, one could purchase a WiMAX modem with a built-in wireless Wi-Fi router, allowing one to connect multiple devices to create a LAN.

Using WiMAX could be an advantage, since it is typically faster than most cable modems with download speeds between 3 and 6 Mbit/s, and generally costs less than cable.

Conformance Testing

TTCN-3 test specification language is used for the purposes of specifying conformance tests for WiMAX implementations. The WiMAX test suite is being developed by a Specialist Task Force at ETSI (STF 252).

Associations

WiMAX Forum

The WiMAX Forum is a non profit organization formed to promote the adoption of WiMAX compatible products and services.

A major role for the organization is to certify the interoperability of WiMAX products. Those that pass conformance and interoperability testing achieve the "WiMAX Forum Certified" designation, and can display this mark on their products and marketing materials. Some vendors claim that their equipment is "WiMAX-ready", "WiMAX-compliant", or "pre-WiMAX", if they are not officially WiMAX Forum Certified.

Another role of the WiMAX Forum is to promote the spread of knowledge about WiMAX. In order to do so, it has a certified training program that is currently offered in English and French. It also offers a series of member events and endorses some industry events.

WiSOA logo

WiMAX Spectrum Owners Alliance

WiSOA was the first global organization composed exclusively of owners of WiMAX spectrum with plans to deploy WiMAX technology in those bands. WiSOA focused on the regulation, commercialisation, and deployment of WiMAX spectrum in the 2.3–2.5 GHz and the 3.4–3.5 GHz ranges. WiSOA merged with the Wireless Broadband Alliance in April 2008.

Telecommunications Industry Association

In 2011, the Telecommunications Industry Association released three technical standards (TIA-1164, TIA-1143, and TIA-1140) that cover the air interface and core networking aspects of Wi-Max High-Rate Packet Data (HRPD) systems using a Mobile Station/Access Terminal (MS/AT) with a single transmitter.

Competing Technologies

Within the marketplace, WiMAX's main competition came from existing, wide-ly deployed wireless systems such as Universal Mobile Telecommunications System (UMTS), CDMA2000, existing Wi-Fi and mesh networking.

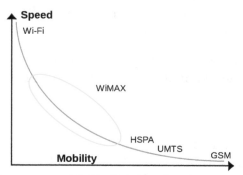

Speed vs. mobility of wireless systems: Wi-Fi, High Speed Packet Access (HSPA), Universal Mobile Telecommunications System (UMTS), GSM

In the future, competition will be from the evolution of the major cellular standards to 4G, high-bandwidth, low-latency, all-IP networks with voice services built on top. The worldwide move to 4G for GSM/UMTS and AMPS/TIA (including CDMA2000) is the 3GPP Long Term Evolution (LTE) effort.

The LTE Standard was finalized in December 2008, with the first commercial de-ployment of LTE carried out by TeliaSonera in Oslo and Stockholm in December, 2009. Since then, LTE has seen increasing adoption by mobile carriers around the world.

In some areas of the world, the wide availability of UMTS and a general desire for stan-dardization has meant spectrum has not been allocated for WiMAX: in July 2005, the EU-wide frequency allocation for WiMAX was blocked.

Harmonization

Early WirelessMAN standards, The European standard HiperMAN and Korean stan-dard WiBro were harmonized as part of WiMAX and are no longer seen as competition but as complementary. All networks now being deployed in South Korea, the home of the WiBro standard, are now WiMAX.

Comparison With Other Mobile Internet Standards

The following table only shows peak rates which are potentially very misleading. In addition, the comparisons listed are not normalized by physical channel size (i.e., spec-trum used to achieve the listed peak rates); this obfuscates spectral efficiency and net through-put capabilities of the different wireless technologies listed below.

Notes: All speeds are theoretical maximums and will vary by a number of factors, including the use of external antennas, distance from the tower and the ground speed (e.g. communications on a train may be poorer than when standing still). Usually the bandwidth is shared between several terminals. The performance of each technology is determined by a number of constraints, including the spectral efficiency of the technology, the cell sizes used, and the amount of spectrum available. For more information, *Comparison of wireless data standards*.

For more comparison tables, see bit rate progress trends, comparison of mobile phone standards, spectral efficiency comparison table and OFDM system comparison table.

Development

The IEEE 802.16m-2011 standard was the core technology for WiMAX 2. The IEEE 802.16m standard was submitted to the ITU for IMT-Advanced standardization. IEEE 802.16m is one of the major candidates for IMT-Advanced technologies by ITU. Among many enhancements, IEEE 802.16m systems can provide four times faster data speed than the WiMAX Release 1.

WiMAX Release 2 provided backward compatibility with Release 1. WiMAX operators could migrate from release 1 to release 2 by upgrading channel cards or software. The WiMAX 2 Collaboration Initiative was formed to help this transition.

It was anticipated that using 4X2 MIMO in the urban microcell scenario with only a single 20 MHz TDD channel available system wide, the 802.16m system can support both 120 Mbit/s downlink and 60 Mbit/s uplink per site simultaneously. It was expected that the WiMAX Release 2 would be available commercially in the 2011–2012 timeframe.

Interference

A field test conducted in 2007 by SUIRG (Satellite Users Interference Reduction Group) with support from the U.S. Navy, the Global VSAT Forum, and several member organizations yielded results showing interference at 12 km when using the same channels for both the WiMAX systems and satellites in C-band.

Deployments

As of October 2010, the WiMAX Forum claimed over 592 WiMAX (fixed and mobile) networks deployed in over 148 countries, covering over 621 million subscribers. By February 2011, the WiMAX Forum cited coverage of over 823 million people, and estimate over 1 billion subscribers by the end of the year.

South Korea launched a WiMAX network in the 2nd quarter of 2006. By the end of 2008 there were 350,000 WiMAX subscribers in Korea.

Worldwide, by early 2010 WiMAX seemed to be ramping quickly relative to other available technologies, though access in North America lagged. Yota, the largest WiMAX network operator in the world in 4Q 2009, announced in May 2010 that it will move new network deployments to LTE and, subsequently, change its existing networks as well.

A study published September 2010 by Blycroft Publishing estimated 800 management contracts from 364 WiMAX operations worldwide offering active services (launched or still trading as opposed to just licensed and still to launch).

References

- K. Fazel and S. Kaiser, Multi-Carrier and Spread Spectrum Systems: From OFDM and MC-CDMA to LTE and WiMAX, 2nd Edition, John Wiley & Sons, 2008, ISBN 978-0-470-99821-2

- M. Ergen, Mobile Broadband - Including WiMAX and LTE, Springer, NY, 2009 ISBN 978-0-387-68189-4

Wireless Technology: An Integrated Study

Gone are the days of wired devices that were cumbersome and lacked mobility. The common wireless technologies in use nowadays are Bluetooth, Wi-Fi and mobile phones. This chapter discusses wireless technology, wireless network and wireless mesh network. A section is committed to examining the comparison of wireless data standards. The topics discussed in the chapter are of great importance to broaden the existing knowledge on wireless technology.

Wireless

A handheld On-board communication station of the maritime mobile service

Wireless communication is the transfer of information or power between two or more points that are not connected by an electrical conductor. The most common wireless technologies use radio. With radio waves distances can be short, such as a few meters for television or as far as thousands or even millions of kilometers for deep-space radio communications. It encompasses various types of fixed, mobile, and portable applications, including two-way radios, cellular telephones, personal digital assistants (PDAs), and wireless networking. Other examples of applications of radio *wireless technology* include GPS units, garage door openers, wireless computer mice, keyboards and head-sets, headphones, radio receivers, satellite television, broadcast television and cordless telephones.

Somewhat less common methods of achieving wireless communications include the use of other electromagnetic wireless technologies, such as light, magnetic, or electric fields or the use of sound. The term *wireless* has been used twice in communications history, with slightly different meaning. It was initially used from about 1890 for the first radio transmitting and receiving technology, as in *wireless telegraphy*, until the new word *radio* replaced it around 1920. The term was revived in the 1980s and 1990s mainly to distinguish digital devices that communicate without wires, such as the ex-amples listed in the previous paragraph, from those that require wires or cables. This is its primary usage in the 2000s. LTE, LTE-Advanced, Wi-Fi and Bluetooth are common modern wireless technologies used in the 2000s.

Wireless operations permit services, such as long-range communications, that are im-possible or impractical to implement with the use of wires. The term is commonly used in the telecommunications industry to refer to telecommunications systems (e.g. radio transmitters and receivers, remote controls, etc.) which use some form of energy (e.g. radio waves, acoustic energy,) to transfer information without the use of wires. Infor-mation is transferred in this manner over both short and long distances.

History

Photophone

Bell and Tainter's photophone, of 1880.

The world's first wireless telephone conversation occurred in 1880, when Alexander Graham Bell and Charles Sumner Tainter invented and patented the photophone, a telephone that conducted audio conversations wirelessly over modulated light beams (which are narrow projections of electromagnetic waves). In that distant era, when util-ities did not yet exist to provide electricity and lasers had not even been imagined in science fiction, there were no practical applications for their invention, which was high-ly limited by the availability of both sunlight and good weather. Similar to free-space optical communication, the photophone also required a clear line of sight between its transmitter and its receiver. It would be several decades before the photophone's principles found their first practical applications in military communications and later in fiber-optic communications.

Early Usage

David E. Hughes transmitted radio signals over a few hundred yards using a clockwork keyed transmitter in 1878. As this was before Maxwell's work was understood, Hughes' contemporaries dismissed his achievement as mere "Induction." In 1885, Thomas Edison used a vibrator magnet for induction transmission. In 1888, Edison deployed a system of signaling on the Lehigh Valley Railroad. In 1891, Edison obtained the wireless patent for this method using inductance (U.S. Patent 465,971).

In 1888, Heinrich Hertz demonstrated the existence of electromagnetic waves, the underlying basis of most wireless technology. The theory of electromagnetic waves was predicted from the research of James Clerk Maxwell and Michael Faraday. Hertz demonstrated that electromagnetic waves traveled through space in straight lines, could be transmitted, and could be received by an experimental apparatus. Hertz did not follow up on the experiments. Jagadish Chandra Bose around this time developed an early wireless detection device and helped increase the knowledge of millimeter-length electromagnetic waves. Later inventors implemented practical applications of wireless radio communication and radio remote control technology.

Radio

Marconi transmitting the first radio signal across the Atlantic.

The term "wireless" came into public use to refer to a radio receiver or transceiver (a dual purpose receiver and transmitter device), establishing its use in the field of wireless telegraphy early on; now the term is used to describe modern wireless connections such as in cellular networks and wireless broadband Internet. It is also used in a general sense to refer to any operation that is implemented without the use of wires, such as "wireless remote control" or "wireless energy transfer", regardless of the specific technology (e.g. radio, infrared, ultrasonic) used. Guglielmo Marconi and Karl Ferdinand Braun were awarded the 1909 Nobel Prize for Physics for their contribution to wireless telegraphy.

Modes

Wireless communications can be via:

Radio

radio communication, microwave communication, for example long-range line-of-sight via highly directional antennas, or short-range communication,

Free-space Optical

An 8-beam free space optics laser link, rated for 1 Gbit/s at a distance of approximately 2 km. The receptor is the large disc in the middle, the transmitters the smaller ones. To the top and right corner a monocular for assisting the alignment of the two heads.

Free-space optical communication (FSO) is an optical communication technology that uses light propagating in free space to transmit wirelessly data for telecommunications or computer networking. "Free space" means the light beams travel through the open air or outer space. This contrasts with other communication technologies that use light beams traveling through transmission lines such as optical fiber or dielectric "light pipes".

The technology is useful where physical connections are impractical due to high costs or other considerations. For example, free space optical links are used in cities between office buildings which are not wired for networking, where the cost of running cable through the building and under the street would be prohibitive. Another widely used example is consumer IR devices such as remote controls and IrDA (Infrared Data Association) networking, which is used as an alternative to WiFi networking to allow laptops, PDAs, printers, and digital cameras to exchange data.

Sonic

Sonic, especially ultrasonic short range communication involves the transmission and reception of sound.

Electromagnetic Induction

Electromagnetic induction short range communication and power. This has been used in biomedical situations such as pacemakers, as well as for short-range Rfid tags.

Services

Common examples of wireless equipment include:

- Infrared and ultrasonic remote control devices

- Professional LMR (Land Mobile Radio) and SMR (Specialized Mobile Radio) typically used by business, industrial and Public Safety entities.

- Consumer Two-way radio including FRS Family Radio Service, GMRS (General Mobile Radio Service) and Citizens band ("CB") radios.

- The Amateur Radio Service (Ham radio).

- Consumer and professional Marine VHF radios.

- Airband and radio navigation equipment used by aviators and air traffic control

- Cellular telephones and pagers: provide connectivity for portable and mobile applications, both personal and business.

- Global Positioning System (GPS): allows drivers of cars and trucks, captains of boats and ships, and pilots of aircraft to ascertain their location anywhere on earth.

- Cordless computer peripherals: the cordless mouse is a common example; wireless headphones, keyboards, and printers can also be linked to a computer via wireless using technology such as Wireless USB or Bluetooth

- Cordless telephone sets: these are limited-range devices, not to be confused with cell phones.

- Satellite television: Is broadcast from satellites in geostationary orbit. Typical services use direct broadcast satellite to provide multiple television channels to viewers.

Computers

- Wi-Fi

- Cordless computer peripherals:

    o mouse

    o headphones,

- keyboards,
- printers,
- USB and,
- Bluetooth
- Wireless networking
  - To span a distance beyond the capabilities of typical cabling,
  - To provide a backup communications link in case of normal network failure,
  - To link portable or temporary workstations,
  - To overcome situations where normal cabling is difficult or financially impractical, or
  - To remotely connect mobile users or networks.

Developers need to consider some parameters involving Wireless RF technology for better developing wireless networks:

- Sub-GHz versus 2.4 GHz frequency trends
- Operating range and battery life
- Sensitivity and data rate
- Network topology and node intelligence

Applications may involve point-to-point communication, point-to-multipoint communication, broadcasting, cellular networks and other wireless networks, Wi-Fi technology.

Cordless

The term "wireless" should not be confused with the term "cordless", which is generally used to refer to powered electrical or electronic devices that are able to operate from a portable power source (e.g., a battery pack) without any cable or cord to limit the mobility of the cordless device through a connection to the mains power supply. Some cordless devices, such as cordless telephones, are also wireless in the sense that information is transferred from the cordless telephone to the phone's base unit via some wireless communications link. This has caused some disparity in the usage of the term "cordless", for example in Digital Enhanced Cordless Telecommunications.

Electromagnetic Spectrum

Light, colors, AM and FM radio, and electronic devices make use of the electromagnetic spectrum. The frequencies of the radio spectrum that are available for use for communication are treated as a public resource and are regulated by national organizations such as the Federal Communications Commission in the USA, or Ofcom in the United Kingdom. This determines which frequency ranges can be used for what purpose and by whom. In the absence of such control or alternative arrangements such as a privatized electromagnetic spectrum, chaos might result if, for example, airlines did not have specific frequencies to work under and an amateur radio operator were interfering with the pilot's ability to land an aircraft. Wireless communication spans the spectrum from 9 kHz to 300 GHz.

Applications

Mobile Telephones

One of the best-known examples of wireless technology is the mobile phone, also known as a cellular phone, with more than 4.6 billion mobile cellular subscriptions worldwide as of the end of 2010. These wireless phones use radio waves from signal-transmission towers to enable their users to make phone calls from many locations worldwide. They can be used within range of the mobile telephone site used to house the equipment required to transmit and receive the radio signals from these instruments.

Data Communications

Wireless data communications are an essential component of mobile computing. The various available technologies differ in local availability, coverage range and performance, and in some circumstances, users must be able to employ multiple connection types and switch between them. To simplify the experience for the user, connection manager software can be used, or a mobile VPN deployed to handle the multiple connections as a secure, single virtual network. Supporting technologies include:

- Wi-Fi is a wireless local area network that enables portable computing devices to connect easily to the Internet. Standardized as IEEE 802.11 a,b,g,n, Wi-Fi approaches speeds of some types of wired Ethernet. Wi-Fi has become the de facto standard for access in private homes, within offices, and at public hotspots. Some businesses charge customers a monthly fee for service, while others have begun offering it for free in an effort to increase the sales of their goods.

- Cellular data service offers coverage within a range of 10-15 miles from the nearest cell site. Speeds have increased as technologies have evolved, from earlier technologies such as GSM, CDMA and GPRS, to 3G networks such as W-CDMA, EDGE or CDMA2000.

- Mobile Satellite Communications may be used where other wireless connections are unavailable, such as in largely rural areas or remote locations. Satellite communications are especially important for transportation, aviation, maritime and military use.

- Wireless Sensor Networks are responsible for sensing noise, interference, and activity in data collection networks. This allows us to detect relevant quantities, monitor and collect data, formulate clear user displays, and to perform decision-making functions

Energy Transfer

Wireless energy transfer is a process whereby electrical energy is transmitted from a power source to an electrical load (Computer Load) that does not have a built-in power source, without the use of interconnecting wires. There are two different fundamental methods for wireless energy transfer. They can be transferred using either far-field methods that involve beaming power/lasers, radio or microwave transmissions or near-field using induction. Both methods utilize electromagnetism and magnetic fields.

Medical Technologies

New wireless technologies, such as mobile body area networks (MBAN), have the capability to monitor blood pressure, heart rate, oxygen level and body temperature. The MBAN works by sending low powered wireless signals to receivers that feed into nursing stations or monitoring sites. This technology helps with the intentional and unintentional risk of infection or disconnection that arise from wired connections.

Computer Interface Devices

Answering the call of customers frustrated with cord clutter, many manufacturers of computer peripherals turned to wireless technology to satisfy their consumer base. Originally these units used bulky, highly local transceivers to mediate between a computer and a keyboard and mouse; however, more recent generations have used small, high-quality devices, some even incorporating Bluetooth. These systems have become so ubiquitous that some users have begun complaining about a lack of wired peripherals. Wireless devices tend to have a slightly slower response time than their wired counterparts; however, the gap is decreasing.A battery powers computer interface devices such as a keyboard or mouse and send signals to a receiver through a USB port by the way of a radio frequency (RF) receiver. The RF design makes it possible for signals to be transmitted wirelessly and expands the range of efficient use, usually up to 10 feet. Distance, physical obstacles, competing signals, and even human bodies can all degrade the signal quality. Concerns about the security of wireless keyboards arose at the end of 2007, when it was revealed that Microsoft's implementation of encryption in some of its 27 MHz models was highly insecure.

Categories of Implementations, Devices and Standards

- Radio station in accordance with ITU RR
- Radiocommunication service in accordance with ITU RR
- Radio communication system
- Land Mobile Radio or Professional Mobile Radio: TETRA, P25, OpenSky, EDACS, DMR, dPMR
- Cordless telephony:DECT (Digital Enhanced Cordless Telecommunications)
- Cellular networks: 0G, 1G, 2G, 3G, Beyond 3G (4G), Future wireless
- List of emerging technologies
- Short-range point-to-point communication : Wireless microphones, Remote controls, IrDA, RFID (Radio Frequency Identification), TransferJet, Wireless USB, DSRC (Dedicated Short Range Communications), EnOcean, Near Field Communication
- Wireless sensor networks: ZigBee, EnOcean; Personal area networks, Bluetooth, TransferJet, Ultra-wideband (UWB from WiMedia Alliance).
- Wireless networks: Wireless LAN (WLAN), (IEEE 802.11 branded as Wi-Fi and HiperLAN), Wireless Metropolitan Area Networks (WMAN) and (LMDS, Wi-MAX, and HiperMAN)

Wireless Network

Wireless icon

A wireless network is any type of computer network that uses wireless data connections for connecting network nodes.

Wireless networking is a method by which homes, telecommunications networks and enterprise (business) installations avoid the costly process of introducing cables into a building, or as a connection between various equipment locations. Wireless telecommunications networks are generally implemented and administered using radio communication. This implementation takes place at the physical level (layer) of the OSI model network structure.

Examples of wireless networks include cell phone networks, Wireless local networks, wireless sensor networks, satellite communication networks, and terrestrial microwave networks.

History

Wireless Links

Computers are very often connected to networks using wireless links

- Terrestrial microwave – Terrestrial microwave communication uses Earth-based transmitters and receivers resembling satellite dishes. Terrestrial microwaves are in the low gigahertz range, which limits all communications to line-of-sight. Relay stations are spaced approximately 48 km (30 mi) apart.

- Communications satellites – Satellites communicate via microwave radio waves, which are not deflected by the Earth's atmosphere. The satellites are stationed in space, typically in geosynchronous orbit 35,400 km (22,000 mi) above the equator. These Earth-orbiting systems are capable of receiving and relaying voice, data, and TV signals.

- Cellular and PCS systems use several radio communications technologies. The systems divide the region covered into multiple geographic areas. Each area has a low-power transmitter or radio relay antenna device to relay calls from one area to the next area.

- Radio and spread spectrum technologies – Wireless local area networks use a high-frequency radio technology similar to digital cellular and a low-frequency radio technology. Wireless LANs use spread spectrum technology to enable communication between multiple devices in a limited area. IEEE 802.11 defines a common flavor of open-standards wireless radio-wave technology known as Wifi.

- Free-space optical communication uses visible or invisible light for communications. In most cases, line-of-sight propagation is used, which limits the physical positioning of communicating devices.

Types of Wireless Networks

Wireless PAN

Wireless personal area networks (WPANs) interconnect devices within a relatively small area, that is generally within a person's reach. For example, both Bluetooth radio and invisible infrared light provides a WPAN for interconnecting a headset to a laptop. ZigBee also supports WPAN applications. Wi-Fi PANs are becoming commonplace (2010) as equipment designers start to integrate Wi-Fi into a variety of consumer electronic devices. Intel "My WiFi" and Windows 7 "virtual Wi-Fi" capabilities have made Wi-Fi PANs simpler and easier to set up and configure.

Wireless LAN

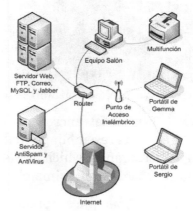

Wireless LANs are often used for connecting to local resources and to the Internet

A wireless local area network (WLAN) links two or more devices over a short distance using a wireless distribution method, usually providing a connection through an access point for internet access. The use of spread-spectrum or OFDM technologies may allow users to move around within a local coverage area, and still remain connected to the network.

Products using the IEEE 802.11 WLAN standards are marketed under the Wi-Fi brand name. Fixed wireless technology implements point-to-point links between computers or networks at two distant locations, often using dedicated microwave or modulated laser light beams over line of sight paths. It is often used in cities to connect networks in two or more buildings without installing a wired link.

Wireless Mesh Network

A wireless mesh network is a wireless network made up of radio nodes organized in a mesh topology. Each node forwards messages on behalf of the other nodes. Mesh networks can "self-heal", automatically re-routing around a node that has lost power.

Wireless MAN

Wireless metropolitan area networks are a type of wireless network that connects several wireless LANs.

- WiMAX is a type of Wireless MAN and is described by the IEEE 802.16 standard.

Wireless WAN

Wireless wide area networks are wireless networks that typically cover large areas, such as between neighbouring towns and cities, or city and suburb. These networks can be used to connect branch offices of business or as a public Internet access system. The wireless connections between access points are usually point to point microwave links using parabolic dishes on the 2.4 GHz band, rather than omni-directional antennas used with smaller networks. A typical system contains base station gateways, access points and wireless bridging relays. Other configurations are mesh systems where each access point acts as a relay also. When combined with renewable energy systems such as photovoltaic solar panels or wind systems they can be stand alone systems.

Global Area Network

A global area network (GAN) is a network used for supporting mobile across an arbitrary number of wireless LANs, satellite coverage areas, etc. The key challenge in mobile communications is handing off user communications from one local coverage area to the next. In IEEE Project 802, this involves a succession of terrestrial wireless LANs.

Space Network

Space networks are networks used for communication between spacecraft, usually in the vicinity of the Earth. The example of this is NASA's Space Network.

Different Uses

Some examples of usage include cellular phones which are part of everyday wireless networks, allowing easy personal communications. Another example, Intercontinental network systems, use radio satellites to communicate across the world. Emergency services such as the police utilize wireless networks to communicate effectively as well. Individuals and businesses use wireless networks to send and share data rapidly, whether it be in a small office building or across the world.

Properties

General

In a general sense, wireless networks offer a vast variety of uses by both business and home users.

"Now, the industry accepts a handful of different wireless technologies. Each wireless technology is defined by a standard that describes unique functions at both the Physical and the Data Link layers of the OSI model. These standards differ in their specified signaling methods, geographic ranges, and frequency usages, among other things. Such differences can make certain technologies better suited to home networks and others better suited to network larger organizations."

Performance

Each standard varies in geographical range, thus making one standard more ideal than the next depending on what it is one is trying to accomplish with a wireless network. The performance of wireless networks satisfies a variety of applications such as voice and video. The use of this technology also gives room for expansions, such as from 2G to 3G and, most recently, 4G technology, which stands for the fourth generation of cell phone mobile communications standards. As wireless networking has become commonplace, sophistication increases through configuration of network hardware and software, and greater capacity to send and receive larger amounts of data, faster, is achieved.

Space

Space is another characteristic of wireless networking. Wireless networks offer many advantages when it comes to difficult-to-wire areas trying to communicate such as across a street or river, a warehouse on the other side of the premises or buildings that are physically separated but operate as one. Wireless networks allow for users to designate a certain space which the network will be able to communicate with other devices through that network. Space is also created in homes as a result of eliminating clutters of wiring. This technology allows for an alternative to installing physical network mediums such as TPs, coaxes, or fiber-optics, which can also be expensive.

Home

For homeowners, wireless technology is an effective option compared to Ethernet for sharing printers, scanners, and high-speed Internet connections. WLANs help save the cost of installation of cable mediums, save time from physical installation, and also creates mobility for devices connected to the network. Wireless networks are simple and require as few as one single wireless access point connected directly to the Internet via a router.

Wireless Network Elements

The telecommunications network at the physical layer also consists of many interconnected wireline network elements (NEs). These NEs can be stand-alone systems or products that are either supplied by a single manufacturer or are assembled by the service provider (user) or system integrator with parts from several different manufacturers.

Wireless NEs are the products and devices used by a wireless carrier to provide support for the backhaul network as well as a mobile switching center (MSC).

Reliable wireless service depends on the network elements at the physical layer to be protected against all operational environments and applications (GR-3171, *Gener-ic Requirements for Network Elements Used in Wireless Networks – Physical Layer Criteria*).

What are especially important are the NEs that are located on the cell tower to the base station (BS) cabinet. The attachment hardware and the positioning of the antenna and associated closures and cables are required to have adequate strength, robustness, corrosion resistance, and resistance against wind, storms, icing, and other weather conditions. Requirements for individual components, such as hardware, cables, connectors, and closures, shall take into consideration the structure to which they are attached.

Difficulties

Interferences

Compared to wired systems, wireless networks are frequently subject to electromagnetic interference. This can be caused by other networks or other types of equipment that generate radio waves that are within, or close, to the radio bands used for communication. Interference can degrade the signal or cause the system to fail.

Absorption and Reflection

Some materials cause absorption of electromagnetic waves, preventing it from reaching the receiver, in other cases, particularly with metallic or conductive materials reflection occurs. This can cause dead zones where no reception is available. Aluminium foiled thermal isolation in modern homes can easily reduce indoor mobile signals by 10 dB frequently leading to complaints about the bad reception of long-distance rural cell signals.

Multipath Fading

In multipath fading two or more different routes taken by the signal, due to reflections, can cause the signal to cancel out at certain locations, and to be stronger in other places (upfade).

Hidden Node Problem

The hidden node problem occurs in some types of network when a node is visible from a wireless access point (AP), but not from other nodes communicating with that AP. This leads to difficulties in media access control.

Shared Resource Problem

The wireless spectrum is a limited resource and shared by all nodes in the range of its transmitters. Bandwidth allocation becomes complex with multiple participating users. Often users are not aware that advertised numbers (e.g., for IEEE 802.11 equipment or LTE networks) are not their capacity, but shared with all other users and thus the individual user rate is far lower. With increasing demand, the capacity crunch is more and more likely to happen. User-in-the-loop (UIL) may be an alternative solution to ever upgrading to newer technologies for over-provisioning.

Capacity

Channel

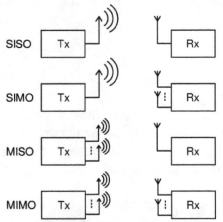

Understanding of SISO, SIMO, MISO and MIMO. Using multiple antennas and transmitting in different frequency channels can reduce fading, and can greatly increase the system capacity.

Shannon's theorem can describe the maximum data rate of any single wireless link, which relates to the bandwidth in hertz and to the noise on the channel.

One can greatly increase channel capacity by using MIMO techniques, where multiple aerials or multiple frequencies can exploit multiple paths to the receiver to achieve much higher throughput – by a factor of the product of the frequency and aerial diversity at each end.

Under Linux, the Central Regulatory Domain Agent (CRDA) controls the setting of channels.

Network

The total network bandwidth depends on how dispersive the medium is (more dispersive medium generally has better total bandwidth because it minimises interference), how many frequencies are available, how noisy those frequencies are, how many aerials are used and whether a directional antenna is in use, whether nodes employ power control and so on.

Cellular wireless networks generally have good capacity, due to their use of directional aerials, and their ability to reuse radio channels in non-adjacent cells. Additionally, cells can be made very small using low power transmitters this is used in cities to give network capacity that scales linearly with population density.

Safety

Wireless access points are also often close to humans, but the drop off in power over distance is fast, following the inverse-square law. The position of the United Kingdom's Health Protection Agency (HPA) is that "...radio frequency (RF) exposures from WiFi are likely to be lower than those from mobile phones." It also saw "...no reason why schools and others should not use WiFi equipment." In October 2007, the HPA launched a new "systematic" study into the effects of WiFi networks on behalf of the UK government, in order to calm fears that had appeared in the media in a recent period up to that time". Dr Michael Clark, of the HPA, says published research on mobile phones and masts does not add up to an indictment of WiFi.

Wireless Mesh Network

A wireless mesh network (WMN) is a communications network made up of radio nodes organized in a mesh topology. It is also a form of wireless ad hoc network. Wireless mesh networks often consist of mesh clients, mesh routers and gateways. The mesh clients are often laptops, cell phones and other wireless devices while the mesh routers forward traffic to and from the gateways which may, but need not, be connected to the Internet. The coverage area of the radio nodes working as a single network is sometimes called a mesh cloud. Access to this mesh cloud is dependent on the radio nodes working in harmony with each other to create a radio network.

A mesh network is reliable and offers redundancy. When one node can no longer operate, the rest of the nodes can still communicate with each other, directly or through one or more intermediate nodes. Wireless mesh networks can self form and self heal. Wireless mesh networks can be implemented with various wireless technologies including 802.11, 802.15, 802.16, cellular technologies and need not be restricted to any one technology or protocol.

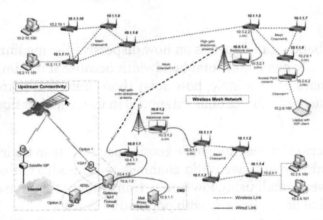

Diagram showing a possible configuration for a wireless mesh network, connected upstream via a VSAT link

History

Architecture

Wireless mesh architecture is a first step towards providing cost effective and dynamic high-bandwidth networks over a specific coverage area. Wireless mesh infrastructure is, in effect, a network of routers minus the cabling between nodes. It's built of peer radio devices that don't have to be cabled to a wired port like traditional WLAN access points (AP) do. Mesh infrastructure carries data over large distances by splitting the distance into a series of short hops. Intermediate nodes not only boost the signal, but cooperatively pass data from point A to point B by making forwarding decisions based on their knowledge of the network, i.e. perform routing. Such an architecture may, with careful design, provide high bandwidth, spectral efficiency, and economic advantage over the coverage area.

Wireless mesh networks have a relatively stable topology except for the occasional failure of nodes or addition of new nodes. The path of traffic, being aggregated from a large number of end users, changes infrequently. Practically all the traffic in an infrastructure mesh network is either forwarded to or from a gateway, while in ad hoc networks or client mesh networks the traffic flows between arbitrary pairs of nodes.

Management

This type of infrastructure can be decentralized (with no central server) or centrally managed (with a central server). Both are relatively inexpensive, and can be very reliable and resilient, as each node needs only transmit as far as the next node. Nodes act as routers to transmit data from nearby nodes to peers that are too far away to reach in a single hop, resulting in a network that can span larger distances. The topology of a mesh network is also reliable, as each node is connected to several other nodes. If one node drops out of the network, due to hardware failure or any other reason, its neighbors can quickly find another route using a routing protocol.

Applications

Mesh networks may involve either fixed or mobile devices. The solutions are as diverse as communication needs, for example in difficult environments such as emergency situations, tunnels, oil rigs, battlefield surveillance, high-speed mobile-video applications on board public transport or real-time racing-car telemetry. An important possible application for wireless mesh networks is VoIP. By using a Quality of Service scheme, the wireless mesh may support local telephone calls to be routed through the mesh.

Some current applications:

- U.S. military forces are now using wireless mesh networking to connect their computers, mainly ruggedized laptops, in field operations.

- Electric meters now being deployed on residences transfer their readings from one to another and eventually to the central office for billing without the need for human meter readers or the need to connect the meters with cables.

- The laptops in the One Laptop per Child program use wireless mesh networking to enable students to exchange files and get on the Internet even though they lack wired or cell phone or other physical connections in their area.

- The 66-satellite Iridium constellation operates as a mesh network, with wireless links between adjacent satellites. Calls between two satellite phones are routed through the mesh, from one satellite to another across the constellation, without having to go through an earth station. This makes for a smaller travel distance for the signal, reducing latency, and also allows for the constellation to operate with far fewer earth stations than would be required for 66 traditional communications satellites.

Operation

The principle is similar to the way packets travel around the wired Internet – data will hop from one device to another until it eventually reaches its destination. Dynamic routing algorithms implemented in each device allow this to happen. To implement such dynamic routing protocols, each device needs to communicate routing information to other devices in the network. Each device then determines what to do with the data it receives – either pass it on to the next device or keep it, depending on the protocol. The routing algorithm used should attempt to always ensure that the data takes the most appropriate (fastest) route to its destination.

Multi-radio Mesh

Multi-radio mesh refers to a unique pair of dedicated radios on each end of the link. This means there is a unique frequency used for each wireless hop and thus a dedicated CSMA collision domain. This is a true mesh link where you can achieve maximum performance without bandwidth degradation in the mesh and without adding latency. Thus voice and video applications work just as they would on a wired Ethernet network. In true 802.11 networks, there is no concept of a mesh. There are only APs and Stations. A multi-radio wireless mesh node will dedicate one of the radios to act as a station, and connect to a neighbor node AP radio.

Research Topics

One of the more often cited papers on Wireless Mesh Networks identified the following areas as open research problems in 2005

- New modulation scheme

    o In order to achieve higher transmission rate, new wideband transmission schemes other than OFDM and UWB are needed.

- Advanced antenna processing

    o Advanced antenna processing including directional, smart and multiple antenna technologies is further investigated, since their complexity and cost are still too high for wide commercialization.

- Flexible spectrum management

    o Tremendous efforts on research of frequency-agile techniques are being performed for increased efficiency.

- Cross-layer optimization

    o Cross-layer research is a popular current research topic where information is shared between different communications layers in order to increase the knowledge and current state of the network. This could enable new and more efficient protocols to be developed. A joint protocol which combines various design problems like routing, scheduling, channel assignment etc. can achieve higher performance since it is proven that these problems are strongly co-related. It is important to note that careless cross-layer design could lead to code which is difficult to maintain and extend.

- Software-defined wireless networking

    o Centralized, distributed, or hybrid? - In a new SDN architecture for WDNs is explored that eliminates the need for multi-hop flooding of route information and therefore enables WDNs to easily expand. The key idea is to split network control and data forwarding by using two separate frequency bands. The forwarding nodes and the SDN controller exchange link-state information and other network control signaling in one of the bands, while actual data forwarding takes place in the other band.

- Security

    o A WMN can be seen as a group of nodes (clients or routers) that cooperate to provide connectivity. Such an open architecture, where clients serve as routers to forward data packets, is exposed to many types of attacks that can interrupt the whole network and cause denial of service (DoS) or Distributed Denial of Service (DDoS).

Protocols

Routing Protocols

There are more than 70 competing schemes for routing packets across mesh networks. Some of these include:

- AODV (Ad hoc On-Demand Distance Vector)

- B.A.T.M.A.N. (Better Approach To Mobile Adhoc Networking)

- Babel (protocol) (a distance-vector routing protocol for IPv6 and IPv4 with fast convergence properties)

- DNVR (Dynamic NIx-Vector Routing)

- DSDV (Destination-Sequenced Distance-Vector Routing)

- DSR (Dynamic Source Routing)

- HSLS (Hazy-Sighted Link State)

- HWMP (Hybrid Wireless Mesh Protocol)

- IWMP (Infrastructure Wireless Mesh Protocol) for Infrastructure Mesh Networks by GRECO UFPB-Brazil

- Wireless mesh networks routing protocol (MRP) by Jangeun Jun and Mihail L. Sichitiu

- OLSR (Optimized Link State Routing protocol)

- OORP (OrderOne Routing Protocol) (OrderOne Networks Routing Protocol)
- OSPF (Open Shortest Path First Routing)
- Routing Protocol for Low-Power and Lossy Networks (IETF ROLL RPL protocol, RFC 6550)
- PWRP (Predictive Wireless Routing Protocol)
- TORA (Temporally-Ordered Routing Algorithm)
- ZRP (Zone Routing Protocol)

The IEEE is developing a set of standards under the title 802.11s to define an architecture and protocol for ESS Mesh Networking.

A less thorough list can be found at Ad hoc routing protocol list.

Autoconfiguration Protocols

Standard autoconfiguration protocols, such as DHCP or IPv6 stateless autoconfiguration may be used over mesh networks.

Mesh network specific autoconfiguration protocols include:

- Ad Hoc Configuration Protocol (AHCP)
- Proactive Autoconfiguration (Proactive Autoconfiguration Protocol)
- Dynamic WMN Configuration Protocol (DWCP)

Comparison of Wireless Data Standards

A wide variety of different wireless data technologies exist, some in direct competition with one another, others designed for specific applications. Wireless technologies can be evaluated by a variety of different metrics of which some are described in this entry.

Standards can be grouped as follows in increasing range order:

Personal Area Network (PAN) systems are intended for short range communication between devices typically controlled by a single person. Some examples include wireless headsets for mobile phones or wireless heart rate sensors communicating with a wrist watch. Some of these technologies include standards such as ANT UWB, Bluetooth, ZigBee, and Wireless USB.

Wireless Sensor Networks (WSN / WSAN) are, generically, networks of low-power, low-cost devices that interconnect wirelessly to collect, exchange, and sometimes act-on data collected from their physical environments - "sensor networks". Nodes typically connect in a star or mesh topology. While most individual nodes in a WSAN are expected to have limited range (Bluetooth, ZigBee, 6LoWPAN, etc.), particular nodes may be capable of more expansive communications (Wi-Fi, Cellular networks, etc.) and any individual WSAN can span a wide geographical range. An example of a WSAN would be a collection of sensors arranged throughout an agricultural facility to monitor soil moisture levels, report the data back to a computer in the main office for analysis and trend modeling, and maybe turn on automatic watering spigots if the level is too low.

For wider area communications, Wireless Local Area Network (WLAN) is used. WLANs are often known by their commercial product name Wi-Fi. These systems are used to provide wireless access to other systems on the local network such as other computers, shared printers, and other such devices or even the internet. Typically a WLAN offers much better speeds and delays within the local network than an average consumer's Internet access. Older systems that provide WLAN functionality include DECT and HIPERLAN. These however are no longer in widespread use. One typical characteristic of WLANs is that they are mostly very local, without the capability of seamless movement from one network to another.

Cellular networks or WAN are designed for city-wide/national/global coverage areas and seamless mobility from one access point (often defined as a Base Station) to another allowing seamless coverage for very wide areas. Cellular network technologies are often split into 2nd generation 2G, 3G and 4G networks. Originally 2G networks were voice centric or even voice only digital cellular systems (as opposed to the analog 1G networks). Typical 2G standards include GSM and IS-95 with extensions via GPRS, EDGE and 1xRTT, providing Internet access to users of originally voice centric 2G networks. Both EDGE and 1xRTT are 3G standards, as defined by the ITU, but are usually marketed as 2.9G due to their comparatively low speeds and high delays when compared to true 3G technologies.

True 3G systems such as EV-DO, W-CDMA (including HSPA) provide combined circuit switched and packet switched data and voice services from the outset, usually at far better data rates than 2G networks with their extensions. All of these services can be used to provide combined mobile voice access and Internet access at remote locations.

4G networks provide even higher bitrates and many architectural improvements, which are not necessarily visible to the consumer. The current 4G systems that are deployed widely are HSPA+, WIMAX and LTE. The latter two are pure packet based networks without traditional voice circuit capabilities. These networks provide voice services via VoIP.

Some systems are designed for point-to-point line-of-sight communications, once two such nodes get too far apart they can no longer communicate. Other systems are designed to form a wireless mesh network using one of a variety of routing protocols. In a mesh network, when nodes get too far apart to communicate directly, they can still communicate indirectly through intermediate nodes.

Standards

The following standards are included in this comparison.

Wireless Wide Area Network (WWAN)

- EDGE
- EV-DO x1 Rev 0, Rev A, Rev B and x3 standards.
- Flash-OFDM: FLASH(Fast Low-latency Access with Seamless Handoff)-OFDM (Orthogonal Frequency Division Multiplexing)
- GPRS
- HSPA D and U standards.
- Lorawan
- LTE
- RTT
- UMTS over W-CDMA
- UMTS-TDD
- WiMAX: 802.16 standard

Wireless Local Area Network (WLAN)

- Wi-Fi: 802.11a, 802.11b, 802.11g, 802.11n, 802.11ac standards.

Wireless Personal Area Network (WPAN) and most Wireless Sensor Actor Networks (WSAN)

- 6LoWPAN
- Bluetooth V4.0 with standard protocol and with low energy protocol
- IEEE 802.15.4-2006 (low-level protocol definitions corresponding to the OSI model physical and link layers. ZigBee, 6LoWPAN, etc. build upward in the protocol stack and correspond to the network and transport layers.)
- Thread (network protocol)
- UWB

- Wireless USB

- ZigBee

Overview

Notes: All speeds are theoretical maximums and will vary by a number of factors, including the use of external antennas, distance from the tower and the ground speed (e.g. communications on a train may be poorer than when standing still). Usually the bandwidth is shared between several terminals. The performance of each technology is determined by a number of constraints, including the spectral efficiency of the technology, the cell sizes used, and the amount of spectrum available.

For more comparison tables, see bit rate progress trends, comparison of mobile phone standards, spectral efficiency comparison table and OFDM system comparison table.

Peak Bit Rate and Throughput

When discussing throughput, there is often a distinction between the peak data rate of the physical layer, the theoretical maximum data throughput and typical throughput.

The peak bit rate of the standard is the net bit rate provided by the physical layer in the fastest transmission mode (using the fastest modulation scheme and error code), excluding forward error correction coding and other physical layer overhead.

The theoretical maximum throughput for end user is clearly lower than the peak data rate due to higher layer overheads. Even this is never possible to achieve unless the test is done under perfect laboratory conditions.

The typical throughput is what users have experienced most of the time when well within the usable range to the base station. The typical throughput is hard to measure, and depends on many protocol issues such as transmission schemes (slower schemes are used at longer distance from the access point due to better redundancy), packet retransmissions and packet size. The typical throughput is often even lower because of other traffic sharing the same network or cell, interference or even the fixed line capacity from the base station onwards being limited.

Note that these figures cannot be used to predict the performance of any given standard in any given environment, but rather as benchmarks against which actual experience might be compared.

- Downlink is the throughput from the base station to the user handset or computer.

- Uplink is the throughput from the user handset or computer to the base station.

- Range is the maximum range possible to receive data at 25% of the typical rate.

References

- G. Miao, J. Zander, K-W Sung, and B. Slimane, Fundamentals of Mobile Data Networks, Cambridge University Press, ISBN 1107143217, 2016.

- Dean Tamara (2010). Network+ Guide to Networks (5th ed.). Boston: Cengage Learning. ISBN 978-1-4239-0245-4.

- Jones, George. "Future Proof. How Wireless Energy Transfer Will Kill the Power Cable." MaximumPC. 14 Sept. 2010. Web. 26 Oct. 2013.

- Vilorio, Dennis. "You're a what? Tower Climber" (PDF). Occupational Outlook Quarterly. Archived (PDF) from the original on February 3, 2013. Retrieved December 6, 2013.

- Moser, Max; Schrödel, Philipp (2007-12-05). "27Mhz Wireless Keyboard Analysis Report aka "We know what you typed last summer"" (PDF). Retrieved 6 February 2012.

- "WiMAX and the IEEE 802.16m Air Interface Standard" (PDF). WiMax Forum. 4 April 2010. Retrieved 2012-02-07.

- "High Speed Internet on the Road". Archived from the original on September 3, 2011. Retrieved September 6, 2011.

- "Wi-Fi Personal Area Networks get a boost with Windows 7 and Intel My WiFi". Retrieved 27 April 2010.

Key Concepts of WiMAX Technology

To provide a better understanding of WiMAX technology, it is of utmost importance that key concepts like Yota Egg, Digital subscriber line, IEEE 802.16 and WiMAX MIMO be understood by the reader. These devices provide WiMAX capability in various ways. This chapter discusses these key concepts with technical specifications, discerning characteristics and definitions.

Yota Egg

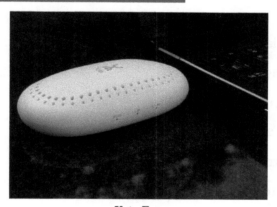

Yota Egg

Yota Egg (also called *Interbro iWWR-1000R*) — an autonomous mobile gateway between WiMax and WiFi wireless networks.

Yota Egg is a battery-powered mobile WiFi access point that connects to the IEEE 802.16e wireless networks (Mobile WiMax), designed to provide Internet access to the devices without a WiMax adapter. With it any 802.11b/g WiFi-enabled device (laptops, communicators, players, camcorders, game consoles and so on) can connect to the WiMax network, regardless of hardware and software. Multiple devices or computers can be connected to the Internet thru one gateway, sharing a common sub-network between them.

This gateway allows you to combine the advantages of WiMax and the wide availability of WiFi adapter. Advantages of this device are its compactness, high autonomy and the ability to quickly deploy a network with high-speed internet access using the popular Wi-Fi standard in the WiMax coverage area. This opens up possibilities of providing a network access in various unsuitable locations and facilities (lack of electricity, poor signal reception inside the building, etc.).

Specifications

Battery time capacity (when fully charged):

- At low load: up to 8 hours
- At high load: up to 4 hours

WiMAX:

- Network standard: IEEE 802.16e-2005
- Frequency range: 2,5-2,7 GHz
- Signal power: 200 mW

Wi-Fi:

- Network Standard: IEEE 802.11b/g
- Frequency Range: 2.4 GHz
- Signal power: 3 mW
- Battery: 3.7 V, 2800 mAh
- Antenna: Internal antenna
- Dimensions: 110 × 61,8 × 28,3 mm
- Weight: 130 g

"Egg" can be powered from USB ports. The current consumption of this device is 2 A, which exceeds the maximum permissible current for USB ports - 500 mA. In order reduce the load on the host port, the cable has two USB plugs on the end, intended for connecting in two USB-ports simultaneously to share the load. The device can be charged from a single Usb Port, But The Charge Rate Will Be Significantly Reduced.

Digital Subscriber Line

Digital subscriber line (*DSL*; originally *digital subscriber loop*) is a family of technologies that are used to transmit digital data over telephone lines. In telecommunications marketing, the term DSL is widely understood to mean asymmetric digital subscriber line (ADSL), the most commonly installed DSL technology, for Internet access. DSL service can be delivered simultaneously with wired telephone service on the same tele-phone line. This is possible because DSL uses higher frequency bands for data. On the customer premises, a DSL filter on each non-DSL outlet blocks any high-frequency interference to enable simultaneous use of the voice and DSL services.

The bit rate of consumer DSL services typically ranges from 256 kbit/s to over 100 Mbit/s in the direction to the customer (downstream), depending on DSL technol-ogy, line conditions, and service-level implementation. Bit rates of 1 Gbit/s have been reached in trials, but most homes are likely to be limited to 500-800 Mbit/s. In ADSL, the data throughput in the upstream direction (the direction to the service provider) is lower, hence the designation of *asymmetric* service. In symmetric digital subscriber line (SDSL) services, the downstream and upstream data rates are equal. Research-ers at Bell Labs have reached speeds of 10 Gbit/s, while delivering 1 Gbit/s symmetri-cal broadband access services using traditional copper telephone lines. These higher speeds are lab results, however. A 2012 survey found that "DSL continues to be the dominant technology for broadband access" with 364.1 million subscribers worldwide.

History

For a long time it was thought that it was not possible to operate a conventional phone-line beyond low-speed limits (typically under 9600 bit/s). In the 1950s, ordinary twist-ed-pair telephone-cable often carried four megahertz (MHz) television signals between studios, suggesting that such lines would allow transmitting many megabits per sec-ond. One such circuit in the UK ran some ten miles (16 km) between the BBC studios in Newcastle-upon-Tyne and the Pontop Pike transmitting station. It was able to give the studios a low quality cue feed but not one suitable for transmission. However, these cables had other impairments besides Gaussian noise, preventing such rates from be-coming practical in the field.

The 1980s saw the development of techniques for broadband communications that al-lowed the limit to be greatly extended. A patent was filed in 1979 for the use of existing telephone wires for both telephones and data terminals that were connected to a re-mote computer via a digital data carrier system. The motivation for digital subscrib-er line technology was the Integrated Services Digital Network (ISDN) specification proposed in 1984 by the CCITT (now ITU-T) as part of Recommendation I.120, later reused as ISDN Digital Subscriber Line (IDSL). Employees at Bellcore (now Telcordia Technologies) developed Asymmetric Digital Subscriber Line (ADSL) by placing wide-band digital signals above the existing baseband analog voice signal carried between telephone company telephone exchanges and customers on conventional twisted pair cabling facilities, and filed a patent in 1988.

Joseph W. Lechleider's contribution to DSL was his insight that an asymmetric ar-rangement offered more than double the bandwidth capacity of symmetric DSL. This allowed Internet service providers to offer efficient service to consumers, who benefited greatly from the ability to download large amounts of data but rarely needed to upload comparable amounts. ADSL supports two modes of transport—fast channel and interleaved channel. Fast channel is preferred for streaming multimedia, where an occasional *dropped bit* is acceptable, but lags are less so. Interleaved channel works better for file transfers, where the delivered data must be error-free but latency (time delay) incurred by the retransmission of error-containing packets is acceptable.

Consumer-oriented ADSL was designed to operate on existing lines already conditioned for Basic Rate Interface ISDN services, which itself is a digital circuit switching service (non-IP), though most incumbent local exchange carriers (ILECs) provision Rate-Adaptive Digital Subscriber Line (RADSL) to work on virtually any available copper pair facility, whether conditioned for BRI or not. Engineers developed high speed DSL facilities such as High bit rate Digital Subscriber Line (HDSL) and Symmetric Digital Subscriber Line (SDSL) to provision traditional Digital Signal 1 (DS1) services over standard copper pair facilities.

Older ADSL standards delivered 8 Mbit/s to the customer over about 2 km (1.2 mi) of unshielded twisted-pair copper wire. Newer variants improved these rates. Distances greater than 2 km (1.2 mi) significantly reduce the bandwidth usable on the wires, thus reducing the data rate. But ADSL loop extenders increase these distances by repeating the signal, allowing the LEC to deliver DSL speeds to any distance.

DSL SoC

Until the late 1990s, the cost of digital signal processors for DSL was prohibitive. All types of DSL employ highly complex digital signal processing algorithms to overcome the inherent limitations of the existing twisted pair wires. Due to the advancements of very-large-scale integration (VLSI) technology, the cost of the equipment associated with a DSL deployment lowered significantly. The two main pieces of equipment are a digital subscriber line access multiplexer (DSLAM) at one end and a DSL modem at the other end.

A DSL connection can be deployed over existing cable. Such deployment, even including equipment, is much cheaper than installing a new, high-bandwidth fiber-optic cable over the same route and distance. This is true both for ADSL and SDSL variations. The commercial success of DSL and similar technologies largely reflects the advances made in electronics over the decades that have increased performance and reduced costs even while digging trenches in the ground for new cables (copper or fiber optic) remains expensive.

In the case of ADSL, competition in Internet access caused subscription fees to drop significantly over the years, thus making ADSL more economical than dial up access. Telephone companies were pressured into moving to ADSL largely due to competition from cable companies, which use DOCSIS cable modem technology to achieve similar speeds. Demand for high bandwidth applications, such as video and file sharing, also contributed to popularize ADSL technology.

Early DSL service required a dedicated dry loop, but when the U.S. Federal Communications Commission (FCC) required ILECs to lease their lines to competing DSL service providers, shared-line DSL became available. Also known as DSL over Unbundled Network Element, this unbundling of services allows a single subscriber to receive two separate services from two separate providers on one cable pair. The DSL service provider's equipment is co-located in the same central office (telephone exchange) as that of the ILEC supplying the customer's pre-existing voice service. The subscriber's circuit is rewired to interface with hardware supplied by the ILEC which combines a DSL frequency and POTS signals on a single copper pair facility.

By 2012 some carriers in the United States reported that DSL remote terminals with fiber backhaul are replacing older ADSL systems.

Operation

Telephones are connected to the telephone exchange via a local loop, which is a physical pair of wires. The local loop was originally intended mostly for the transmission of speech, encompassing an audio frequency range of 300 to 3400 hertz (voiceband or commercial bandwidth). However, as long-distance trunks were gradually converted from analog to digital operation, the idea of being able to pass data through the local loop (by utilizing frequencies above the voiceband) took hold, ultimately leading to DSL.

The local loop connecting the telephone exchange to most subscribers has the capability of carrying frequencies well beyond the 3.4 kHz upper limit of POTS. Depending on the length and quality of the loop, the upper limit can be tens of megahertz. DSL takes advantage of this unused bandwidth of the local loop by creating 4312.5 Hz wide channels starting between 10 and 100 kHz, depending on how the system is configured. Allocation of channels continues at higher and higher frequencies (up to 1.1 MHz for ADSL) until new channels are deemed unusable. Each channel is evaluated for usability in much the same way an analog modem would on a POTS connection. More usable channels equates to more available bandwidth, which is why distance and line quality are a factor (the higher frequencies used by DSL travel only short distances). The pool of usable channels is then split into two different frequency bands for upstream and downstream traffic, based on a preconfigured ratio. This segregation reduces interference. Once the channel groups have been established, the individual channels are bonded into a pair of virtual circuits, one in each direction. Like analog modems, DSL transceivers constantly monitor the quality of each channel and will add or remove them from service depending on whether they are usable. Once upstream and downstream circuits are established, a subscriber can connect to a service such as an Internet service provider or other network services, like a corporate MPLS network.

The underlying technology of transport across DSL facilities uses high-frequency sinusoidal carrier wave modulation, which is an analog signal transmission. A DSL circuit terminates at each end in a modem which modulates patterns of bits into certain high-frequency impulses for transmission to the opposing modem. Signals received from the far-end modem are demodulated to yield a corresponding bit pattern that the modem retransmits, in digital form, to its interfaced equipment, such as a computer, router, switch, etc.

Unlike traditional dial-up modems, which modulate bits into signals in the 300–3400 Hz baseband (voice service), DSL modems modulate frequencies from 4000 Hz to as high as 4 MHz. This frequency band separation enables DSL service and plain old telephone service (POTS) to coexist on the same copper pair facility. On the subscriber's end of the circuit, inline low-pass DSL filters (splitters) are installed on each telephone to filter the high-frequency signals that would otherwise be heard as hiss, but pass voice frequencies. Conversely, high-pass filters already incorporated in the circuitry of DSL modems filter out voice frequencies. Although ADSL and RADSL modulations do not use the voice-frequency band, nonlinear elements in the phone could otherwise generate audible intermodulation and may impair the operation of the data modem in the absence of high-pass filters.

A DSL modem

Because DSL operates above the 3.4 kHz voice limit, it cannot pass through a load coil, which is an inductive coil that is designed to counteract loss caused by shunt capacitance (capacitance between the two wires of the twisted pair). Load coils are commonly set at regular intervals in lines placed only for POTS. A DSL signal cannot pass through a properly installed and working load coil, while voice service cannot be maintained past a certain distance without such coils. Therefore, some areas that are within range for DSL service are disqualified from eligibility because of load coil placement. Because of this, phone companies endeavor to remove load coils on copper loops that can operate without them, and by conditioning other lines to avoid them through the use of fiber to the neighborhood or node (FTTN).

Most residential and small-office DSL implementations reserve low frequencies for POTS, so that (with suitable filters and/or splitters) the existing voice service continues to operate independent of the DSL service. Thus POTS-based communications, including fax machines and analog modems, can share the wires with DSL. Only one DSL modem can use the subscriber line at a time. The standard way to let multiple computers share a DSL connection uses a router that establishes a connection between the DSL modem and a local Ethernet, Powerline, or Wi-Fi network on the customer's premises.

The theoretical foundations of DSL, like much of communication technology, can be traced back to Claude Shannon's seminal 1948 paper: *A Mathematical Theory of Communication*. Generally, higher bit rate transmissions require a wider frequency band, though the ratio of bit rate to symbol rate and thus to bandwidth are not linear due to significant innovations in digital signal processing and digital modulation methods.

Naked DSL

A naked DSL (also known as standalone or dry loop DSL) is a way of providing DSL services without a PSTN (analogue telephony) service. It is useful when the customer does not need the traditional telephony voice service because voice service is received either on top of the DSL services (usually VoIP) or through another network (mobile telephony).

It is also commonly called a "UNE" (for Unbundled Network Element) in the United States; in Australia it is known as a "ULL" (Unconditioned Local Loop); in Belgium it is known as "Raw Copper" and in Turkey it's known as "Yalın Internet". It started making a comeback in the United States in 2004 when Qwest started offering it, closely followed by Speakeasy. As a result of AT&T's merger with SBC, and Verizon's merger with MCI, those telephone companies have an obligation to offer naked DSL to consumers.

In Turkey, since 2011, telephone companies are obliged to offer naked DSL as a result of consumer pressure to the regulatory bodies, however companies can incur additional fees under various label, such as circuit preparation service (devre hazırlama ücreti) or an additional naked DSL fee (yalın adsl ücreti). Although circuit preparation service fee is one-time, the latter is recurring and can constitute as much as 20% of the monthly bill.

Even without the regulatory mandate, however, many ILECs offered naked DSL to consumers. The number of telephone landlines in the United States dropped from 188 million in 2000 to 115 million in 2010, while the number of cellular subscribers has grown to 277 million (as of 2010). This lack of demand for landline voice services has resulted in the expansion of naked DSL availability.

Naked DSL products are also marketed in some other countries e.g., Australia, New Zealand, and Canada.

Typical Setup

On the customer side, the DSL Transceiver, or ATU-R, or more commonly known as a DSL modem, is hooked up to a phone line. The telephone company connects the other end of the line to a DSLAM, which concentrates a large number of individual DSL connections into a single box. The location of the DSLAM depends on the telco, but it cannot be located too far from the user because of attenuation between the DSLAM and the user's DSL modem. It is common for a few residential blocks to be connected to one DSLAM.

The accompanying figure is a schematic of a simple DSL connection (in blue). The right side shows a DSLAM residing in the telephone company's central office. The left side shows the customer premises equipment with an optional router. This router manages a local area network (LAN) off of which are connected some number of PCs. With many service providers, the customer may opt for a modem which contains a wireless router. This option (within the dashed bubble) often simplifies the connection.

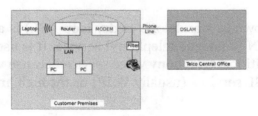

DSL Connection schematic

Example of a DSLAM from 2006

Exchange Equipment

At the exchange, a digital subscriber line access multiplexer (DSLAM) terminates the DSL circuits and aggregates them, where they are handed off to other networking transports. In the case of ADSL, the voice component is also separated at this step, either by a filter integrated in the DSLAM or by a specialized filtering equipment installed before it. The DSLAM terminates all connections and recovers the original digital information.

Customer Equipment

The customer end of the connection consists of a terminal adaptor or "DSL modem". This converts data between the digital signals used by computers and the voltage signal of a suitable frequency range which is then applied to the phone line.

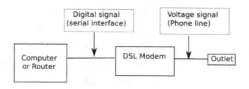

DSL Modem schematic

In some DSL variations (for example, HDSL), the terminal adapter connects directly to the computer via a serial interface, using protocols such as ethernet or V.35. In other cases (particularly ADSL), it is common for the customer equipment to be integrated with higher level functionality, such as routing, firewalling, or other application-specific hardware and software. In this case, the equipment is referred to as a "gateway".

Most DSL technologies require installation of appropriate filters to separate, or "split", the DSL signal from the low-frequency voice signal. The separation can take place either at the demarcation point, or with filters installed at the telephone outlets inside the customer premises. Each way has its practical and economic limitations.

When the DSL modem powers up it goes through a series of steps to establish connections. The actual process varies from modem to modem but generally involves the following steps:

1. The DSL transceiver performs a self-test.

2. The DSL transceiver then attempts to synchronize with the DSLAM. Data can only come into the computer when the DSLAM and the modem are synchronized. The synchronization process is relatively quick (in the range of seconds) but is very complex, involving extensive tests that allow both sides of the con-

nection to optimize the performance according to the characteristics of the line in use. External, or standalone modem units have an indicator labeled "CD", "DSL", or "LINK", which can be used to tell if the modem is synchronized. During synchronization the light flashes; when synchronized, the light stays lit, usually with a green color.

3. The DSL transceiver checks the connection between the DSL transceiver and the computer. For residential variations of DSL, this is usually the Ethernet (RJ-45) port or a USB port; in rare models, a FireWire port is used. Older DSL modems sported a native ATM interface (usually, a 25 Mbit/s serial interface). Also, some variations of DSL (such as SDSL) use synchronous serial connections.

Modern DSL gateways have more functionality and usually go through an initialization procedure very similar to a PC boot up. The system image is loaded from the flash memory; the system boots, synchronizes the DSL connection and establishes the IP connection between the local network and the service provider, using protocols such as DHCP or PPPoE. (According to a 2007 book, the PPPoE method far outweighed DHCP in terms of deployment on DSL lines, and PAP was the predominant form of subscriber authentication used in such circumstances.) The system image can usually be updated to correct bugs, or to add new functionality.

Protocols and Configurations

Many DSL technologies implement an Asynchronous Transfer Mode (ATM) layer over the low-level bitstream layer to enable the adaptation of a number of different technologies over the same link.

DSL implementations may create bridged or routed networks. In a bridged configuration, the group of subscriber computers effectively connect into a single subnet. The earliest implementations used DHCP to provide network details such as the IP address to the subscriber equipment, with authentication via MAC address or an assigned host name. Later implementations often use Point-to-Point Protocol (PPP) to authenticate with a user ID and password, and to provide network details (Point-to-Point Protocol over Ethernet (PPPoE) or Point-to-Point Protocol over ATM (PPPoA)).

Transmission Methods

Transmission methods vary by market, region, carrier, and equipment.

- 2B1Q: Two-binary, one-quaternary, used for IDSL and HDSL

- CAP: Carrierless Amplitude Phase Modulation - deprecated in 1996 for ADSL, used for HDSL

- TC-PAM: Trellis Coded Pulse Amplitude Modulation, used for HDSL2 and SHDSL

- DMT: Discrete multitone modulation, the most common kind, also known as OFDM (Orthogonal frequency-division multiplexing)

DSL Technologies

DSL technologies (sometimes summarized as *xDSL*) include:

- Symmetric digital subscriber line (SDSL), umbrella term for xDSL where the bitrate is equal in both directions.

  - ISDN digital subscriber line (IDSL), ISDN based technology that provides a bitrate equivalent to two ISDN bearer and one data channel, 144 kbit/s symmetric over one pair

  - High bit rate digital subscriber line (HDSL), ITU-T G.991.1, the first DSL technology that used a higher frequency spectrum than ISDN, 1,544 kbit/s and 2,048 kbit/s symmetric services, either on 2 or 3 pairs at 784 kbit/s each, 2 pairs at 1,168 kbit/s each, or one pair at 2,320 kbit/s

  - High bit rate digital subscriber line 2/4 (HDSL2, HDSL4), ANSI, 1,544 kbit/s symmetric over one pair (HDSL2) or two pairs (HDSL4)

  - Symmetric digital subscriber line (SDSL), specific proprietary technology, up to 1,544 kbit/s symmetric over one pair

  - Single-pair high-speed digital subscriber line (G.SHDSL), ITU-T G.991.2, standardized successor of HDSL and proprietary SDSL, up to 5,696 kbit/s per pair, up to four pairs

- Asymmetric digital subscriber line (ADSL), umbrella term for xDSL where the bitrate is greater in one direction than the other.

  - ANSI T1.413 Issue 2, up to 8 Mbit/s and 1 Mbit/s

  - G.dmt, ITU-T G.992.1, up to 10 Mbit/s and 1 Mbit/s

  - G.lite, ITU-T G.992.2, more noise and attenuation resistant than G.dmt, up to 1,536 kbit/s and 512 kbit/s

  - Asymmetric digital subscriber line 2 (ADSL2), ITU-T G.992.3, up to 12 Mbit/s and 3.5 Mbit/s

  - Asymmetric digital subscriber line 2 plus (ADSL2+), ITU-T G.992.5, up to 24 Mbit/s and 3.5 Mbit/s

- o Very-high-bit-rate digital subscriber line (VDSL), ITU-T G.993.1, up to 52 Mbit/s and 16 Mbit/s

- o Very-high-bit-rate digital subscriber line 2 (VDSL2), ITU-T G.993.2, an improved version of VDSL, compatible with ADSL2+, sum of both directions up to 200 Mbit/s. G.vector crosstalk cancelling feature (ITU-T G.993.5) can be used to increase range at a given bitrate, e.g. 100 Mbit/s at up to 500 meters.

- o G.fast, ITU-T G.9700 and G.9701, up to approximately 1 Gbit/s aggregate uplink and downlink at 100m. Approved in December 2014, deployments planned for 2016.

- Bonded DSL Rings (DSL Rings), a shared ring topology at 400 Mbit/s

- Etherloop Ethernet local loop

- High Speed Voice and Data Link

- Internet Protocol subscriber line (IPSL), developed by Rim Semiconductor in 2007, allowed for 40 Mbit/s using 26 AWG copper telephone wire at a 5,500 ft (1,700 m) radius, 26 Mbit/s at a 6,000 ft (1,800 m) radius. The company operated until 2008.

- Rate-adaptive digital subscriber line (RADSL), designed to increase range and noise tolerance by sacrificing up stream speed

- Uni-DSL (Uni digital subscriber line or UDSL), technology developed by Texas Instruments, backwards compatible with all DMT standards

- Frequency Division Vectoring, copper networks working with fiber

The line-length limitations from telephone exchange to subscriber impose severe limits on data transmission rates. Technologies such as VDSL provide very high-speed but short-range links. VDSL is used as a method of delivering "triple play" services (typically implemented in fiber to the curb network architectures).

IEEE 802.16

IEEE 802.16 is a series of wireless broadband standards written by the Institute of Electrical and Electronics Engineers (IEEE). The IEEE Standards Board established a working group in 1999 to develop standards for broadband for wireless metropolitan area networks. The Workgroup is a unit of the IEEE 802 local area network and metropolitan area network standards committee.

Although the 802.16 family of standards is officially called WirelessMAN in IEEE, it has been commercialized under the name "WiMAX" (from "Worldwide Interoperability for Microwave Access") by the WiMAX Forum industry alliance. The Forum promotes and certifies compatibility and interoperability of products based on the IEEE 802.16 standards.

The 802.16e-2005 amendment version was announced as being deployed around the world in 2009. The version *IEEE 802.16-2009* was amended by IEEE 802.16j-2009.

Standards

Projects publish draft and proposed standards with the letter "P" prefixed. Once a standard is ratified and published, that "P" gets dropped and replaced by a trailing dash and suffix year of publication.

Projects

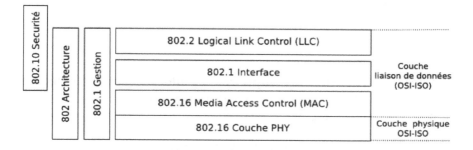

802.16e-2005 Technology

The 802.16 standard essentially standardizes two aspects of the air interface – the physical layer (PHY) and the media access control (MAC) layer. This section provides an overview of the technology employed in these two layers in the mobile 802.16e specification.

PHY

802.16e uses scalable OFDMA to carry data, supporting channel bandwidths of between 1.25 MHz and 20 MHz, with up to 2048 subcarriers. It supports adaptive modulation and coding, so that in conditions of good signal, a highly efficient 64 QAM coding scheme is used, whereas when the signal is poorer, a more robust BPSK coding mechanism is used. In intermediate conditions, 16 QAM and QPSK can also be employed. Other PHY features include support for multiple-input multiple-output (MIMO) antennas in order to provide good non-line-of-sight propagation (NLOS) characteristics (or higher bandwidth) and hybrid automatic repeat request (HARQ) for good error correction performance.

Although the standards allow operation in any band from 2 to 66 GHz, mobile operation is best in the lower bands which are also the most crowded, and therefore most expensive.

MAC

The 802.16 MAC describes a number of *Convergence Sublayers* which describe how wireline technologies such as Ethernet, Asynchronous Transfer Mode (ATM) and Internet Protocol (IP) are encapsulated on the air interface, and how data is classified, etc. It also describes how secure communications are delivered, by using secure key exchange during authentication, and encryption using Advanced Encryption Standard (AES) or Data Encryption Standard (DES) during data transfer. Further features of the MAC layer include power saving mechanisms (using *sleep mode* and *idle mode*) and handover mechanisms.

A key feature of 802.16 is that it is a connection-oriented technology. The subscriber station (SS) cannot transmit data until it has been allocated a channel by the base station (BS). This allows 802.16e to provide strong support for quality of service (QoS).

QoS

Quality of service (QoS) in 802.16e is supported by allocating each connection between the SS and the BS (called a *service flow* in 802.16 terminology) to a specific *QoS class*. In 802.16e, there are 5 QoS classes:

The BS and the SS use a service flow with an appropriate QoS class (plus other parameters, such as bandwidth and delay) to ensure that application data receives QoS treatment appropriate to the application.

Certification

Because the IEEE only sets specifications but does not test equipment for compliance with them, the WiMAX Forum runs a certification program wherein members pay for certification. WiMAX certification by this group is intended to guarantee compliance with the standard and interoperability with equipment from other manufacturers. The mission of the Forum is to promote and certify compatibility and interoperability of broadband wireless products.

WiMAX MIMO

WiMAX MIMO refers to the use of Multiple-input multiple-output communications (MIMO) technology on WiMAX, which is the technology brand name for the implementation of the standard IEEE 802.16.

Background

WiMAX

WiMAX is the technology brand name for the implementation of the standard IEEE 802.16, which specifies the air interface at the PHY (Physical layer) and at the MAC (Medium Access Control layer) . Aside from specifying the support of various channel bandwidths and adaptive modulation and coding, it also specifies the support for MIMO antennas to provide good Non-line-of-sight (NLOS) characteristics.

MIMO

MIMO stands for Multiple Input and Multiple Output, and refers to the technology where there are multiple antennas at the base station and multiple antennas at the mobile device. Typical usage of multiple antenna technology includes cellular phones with two antennas, laptops with two antennas (e.g. built in the left and right side of the screen), as well as CPE devices with multiple sprouting antennas.

The predominant cellular network implementation is to have multiple antennas at the base station and a single antenna on the mobile device. This minimizes the cost of the mobile radio. As the costs for radio frequency (RF) components in mobile devices go down, second antennas in mobile device may become more common. Multiple mobile device antennas are currently used in Wi-Fi technology (e.g. IEEE 802.11n), where WiFi-enabled cellular phones, laptops and other devices often have two or more antennas.

MIMO Technology in WiMAX

WiMAX implementations that use MIMO technology have become important. The use of MIMO technology improves the reception and allows for a better reach and rate of transmission. The implementation of MIMO also gives WiMAX a significant increase in spectral efficiency.

MIMO Auto-negotiation

The 802.16 defined MIMO configuration is negotiated dynamically between each individual base station and mobile station. The 802.16 specification supports the ability to support a mix of mobile stations with different MIMO capabilities. This helps to maximize the sector throughput by leveraging the different capabilities of a diverse set of vendor mobile stations.

Space Time Code

The 802.16 specification supports the Multiple-input and single-output (MISO) technique of Transmit Diversity, which is commonly referred to Space Time Code (STC).

With this method, two or more antennas are employed at the transmitter and one antenna at the receiver. The use of multiple receive antennas (thus MIMO) can further improve the reception of STC transmitted signals.

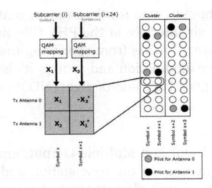

Space Time Code diagram

With a Transmit Diversity rate = 1 (aka "Matrix A" in the 802.16 standard), different data bit constellations are transferred on two different antennas during the same symbol. The conjugate and/or inverse of the same two constellations are transferred again on the same antennas during the next symbol. The data transfer rate with STC remains the same as the baseline case. The received signal is more robust with this method due to the transmission redundancy. This configuration delivers similar performance to the case of two receive antennas and one transmitter antenna.

Spatial Multiplexing

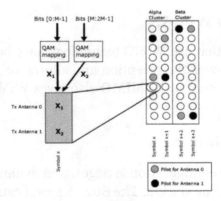

Spatial Multiplexing

The 802.16 specification also supports the MIMO technique of Spatial Multiplexing (SMX), also known as Transmit Diversity rate = 2 (aka "Matrix B" in the 802.16 standard). Instead of transmitting the same bit over two antennas, this method transmits one data bit from the first antenna, and another bit from the second antenna simultaneously, per symbol. As long as the receiver has more than one antenna and the signal is of sufficient quality, the receiver can separate the signals.

This method involves added complexity and expense at both the transmitter and receiver. However, with two transmit antennas and two receive antennas, data can be transmitted twice as fast as compared systems using Space Time Codes with only one receive antenna.

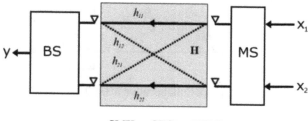

2xSMX or STC+2xMRC

WiMAX Network use of Spatial Multiplexing

One specific use of Spatial Multiplexing is to apply it to users who have the best signal quality, so that less time is spent transmitting to them. Users whose signal quality is too low to allow the spatially multiplexed signals to be resolved stay with conventional transmission. This allows an operator to offer higher data rates to some users and/or to serve more users. The WiMAX specification's dynamic negotiation mechanism helps enable this use.

WiMAX MISO/MIMO with four Antennas

The 802.16 specification also supports the use of four antennas. Three configurations are supported.

WiMAx Four Antenna Mode 1

With rate = 1 using four antennas, data is transmitted four times per symbol, where each time the data is conjugated and/or inverted. This does not change the data rate, but does give the signal more robustness and avoids sudden increases in error rates.

WiMAx Four Antenna Mode 2

With rate = 2 using four antennas, the data rate is only doubled, but increases in robustness since the same data is transmitted twice as compared to only once with using two antennas.

Wimax Four Antenna Matrix C Mode

The third configuration that is only available using four antennas is Matrix C, where a different data bit is transmitted from the four antennas per symbol, which gives it four times the baseline data rate.

Other Advanced MIMO Techniques Applied to WiMAX

Uplink Collaborative MIMO

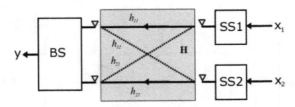

Uplink Collaborative MIMO

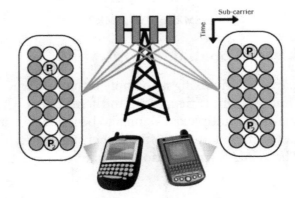

MSs spatially uncorrelated
/Without 3dB power penalty

A related technique is called Uplink Collaborative MIMO, where users transmit at the same time in the same frequency. This type of spatial multiplexing improves the sector throughput without requiring multiple transmit antennas at the mobile device. The common non-MIMO method for this in OFDMA is by scheduling different mobile stations at different points in an OFDMA time-frequency map. Collaborative Spatial Multiplexing (Collaborative MIMO) is comparable to regular spatial multiplexing, where multiple data streams are transmitted from multiple antennas on the same device.

WiMAX Uplink Collaborative MIMO

In the case of WiMAX, Uplink Collaborative MIMO is spatial multiplexing with two different devices, each with one antenna. These transmitting devices are collaborating in the sense that both devices must be synchronized in time and frequency so that the intentional overlapping occurs under controlled circumstances. The two streams of data will then interfere with each other. As long as the signal quality is sufficiently good and the receiver at the base station has at least two antennas, the two data streams can be separated again. This technique is sometimes also termed Virtual Spatial Multiplexing.

Other MIMO-related Radio Techniques Applied to WiMAX

Adaptive Antenna Steering (AAS), a.k.a. Beamforming

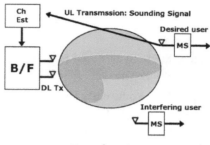

Beamforming

A MIMO-related technique that can be used with WiMAX is called AAS or Beamforming. Multiple antennas and multiple signals are employed, which then shape the beam with the intent of improving transmission to the desired station. The result is reduced interference because the signal going to the desired user is increased and the signal going to other users is reduced.

Cyclic Delay Diversity

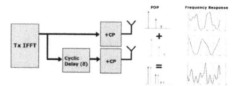

Cyclic Delay Diversity

Another MIMO-related technique that can be used in WiMAX systems, but which is outside of the scope of the 802.16 specification, is known as Cyclic Delay Diversity. In this technique, one or more of the signals are delayed before transmission. Because the signals are coming out of two antennas, their receive spectrums differ as each spectrum is characterized by humps and notches due to multi-path fading. At the receiver the signals combine, which improves reception because the joint reception results in shallower spectral humps and fewer spectral notches. The closer the signal can get towards a flat channel at a certain power level, the higher the throughput that can be obtained.

Radio Conformance Test of WiMAX MIMO

The WiMax Forum has a set of standardized conformance test procedures for PHY and MAC specification compliance called the Radio Conformance Test (RCT). Any technology aspect of a particular implementation of a radio interface must first undergo the RCT. Generally, any aspect of the IEEE 802.16 standard that does not have a test procedure in the RCT may be assumed to not yet be widely implemented.

Silicon Implementations of WiMAX MIMO

Companies that make RFICs that support WiMAX MIMO include Intel, Beceem , NXP Semiconductors and PMC-Sierra.

References

- Philip Golden; Hervé Dedieu; Krista S. Jacobsen (2007). Implementation and Applications of DSL Technology. Taylor & Francis. p. 479. ISBN 978-1-4200-1307-8.

- ARU, ÇAĞDAŞ. "BTK Yalın ADSL Konusunda Düzeltme Yayınladı, Rekabet Kurumu Dünkü Açıklaması Nedeniyle Özür Diledi". Retrieved 15 February 2015.

- "YALIN ADSL VERGİSİZ 8,13 TL OLDU AMA KİMSE BTK KARARINDAN MEMNUN DEĞİL". Telkoder. Retrieved 15 February 2015.

- Spruyt, Paul; Vanhastel, Stefaan (2013-07-04). "The Numbers are in: Vectoring 2.0 Makes G.fast Faster". TechZine. Alcatel Lucent. Retrieved 2014-02-13.

- Hardy, Stephen (2014-10-22). "G.fast ONT available early next year says Alcatel-Lucent". lightwaveonline.com. Retrieved 2014-10-23.

- "IPSL Special Interest Group". consortium web site. 2007. Archived from the original on September 28, 2008. Retrieved September 15, 2011.

- "Rim Semiconductor Company". official web site. Archived from the original on August 24, 2008. Retrieved September 15, 2011.

- "WiMAX™ operators and vendors from around the world announce new deployments, growing commitments at the 2nd Annual WiMAX Forum® Global Congress". News release. WiMAX Forum. June 4, 2009. Retrieved August 20, 2011.

- "IEEE Approves IEEE 802.16m – Advanced Mobile Broadband Wireless Standard". News release. IEEE Standards Association. March 31, 2011. Retrieved August 20, 2011.

- Vos, Esme (October 6, 2009). "Yota Egg: portable mobile WiMAX to Wi-Fi converter". muniwireless.com. Retrieved 6 February 2010.

Various uses of WiMAX Technology

4

WiMAX is faster, efficient, serves remote areas and can support more users than Wi-Fi and this is the reason it is heralded as the technology of the future. Exploiting this potential, services like Internet access, IPTV and triple play have incorporated WiMAX technology and are thus able to provide cheaper, faster and more cost effective transmission at low spectrums. This chapter reviews these services with the objective of helping the reader understand the limitless potentiality of WiMAX technology.

IPTV

Internet Protocol television (IPTV) is a system through which television services are delivered using the Internet protocol suite over a packet-switched network such as a LAN or the Internet, instead of being delivered through traditional terrestrial, satellite signal, and cable television formats. Unlike downloaded media, IPTV offers the ability to stream the media in smaller batches, directly from the source. As a result, a client media player can begin playing the data (such as a movie) before the entire file has been transmitted.... This is known as streaming media.

IPTV services may be classified into three main groups

- Live television, with or without interactivity related to the current TV show;

- Time-shifted television: catch-up TV (replays a TV show that was broadcast hours or days ago), start-over TV (replays the current TV show from its beginning);

- Video on demand (VOD): browse a catalogue of videos, not related to TV programming.

IPTV is distinguished from Internet television by its ongoing standardisation process (e.g., European Telecommunications Standards Institute) and preferential deployment scenarios in subscriber-based telecommunications networks with high-speed access channels into end-user premises via set-top boxes or other customer-premises equipment.

Definition

Historically, many different definitions of IPTV have appeared, including elementary streams over IP networks, transport streams over IP networks and a number of proprietary systems.

One official definition approved by the International Telecommunication Union focus group on IPTV (ITU-T FG IPTV) is:

IPTV is defined as multimedia services such as television/video/audio/text/graphics/ data delivered over IP based networks managed to provide the required level of quality of service and experience, security, interactivity and reliability.

Another more detailed definition of IPTV is the one given by Alliance for Telecommunications Industry Solutions (ATIS) IPTV Exploratory Group on 2005:

IPTV is defined as the secure and reliable delivery to subscribers of entertainment video and related services. These services may include, for example, Live TV, Video On Demand (VOD) and Interactive TV (iTV). These services are delivered across an access agnostic, packet switched network that employs the IP protocol to transport the audio, video and control signals. In contrast to video over the public Internet, with IPTV deployments, network security and performance are tightly managed to ensure a superior entertainment experience, resulting in a compelling business environment for content providers, advertisers and customers alike.

History

The term IPTV first appeared in 1995 with the founding of Precept Software by Judith Estrin and Bill Carrico. Precept developed an Internet video product named *IP/TV*. IP/TV was a multicast backbone (MBONE) compatible Windows and Unix-based application that transmitted single and multi-source audio and video traffic, ranging from low to DVD quality, using both unicast and IP multicast Real-time Transport Protocol (RTP) and Real time control protocol (RTCP). The software was written primarily by Steve Casner, Karl Auerbach, and Cha Chee Kuan. Precept was acquired by Cisco Systems in 1998. Cisco retains the IP/TV trademark.

Internet radio company AudioNet started the first continuous live webcasts with content from WFAA-TV in January 1998 and KCTU-LP on January 10, 1998.

Kingston Communications, a regional telecommunications operator in the UK, launched KIT (Kingston Interactive Television) the brainchild of Matt Child, an IPTV over digital subscriber line (DSL) broadband interactive TV service in September 1999 after conducting various TV and video on demand (VoD) trials. The operator added additional VoD service in October 2001 with Yes TV, a VoD content provider. Kingston was one of the first companies in the world to introduce IPTV and IP VoD over ADSL as a commercial service. The service became the reference for various changes to UK Government regulations and policy on IPTV. In 2006, the KIT service was discontinued, subscribers having declined from a peak of 10,000 to 4,000.

In 1999, NBTel (now known as Bell Aliant) was the first to commercially deploy Internet protocol television over DSL in Canada using the Alcatel 7350 DSLAM and middleware created by iMagic TV (owned by NBTel's parent company Bruncor). The service was marketed under the brand VibeVision in New Brunswick, and later expanded into Nova Scotia in early 2000 after the formation of Aliant. iMagic TV was later sold to Alcatel.

In 2002, Sasktel was the second in Canada to commercially deploy Internet Protocol (IP) video over DSL, using the Lucent Stinger DSL platform.

In 2005, SureWest Communications was the first North American company to offer high-definition television (HDTV) channels over an IPTV service.

In 2003, Total Access Networks Inc launched an IPTV service, consisting of 100 free IPTV stations worldwide.

In 2005, Bredbandsbolaget launched its IPTV service as the first service provider in Sweden. As of January 2009, they are not the biggest supplier any longer; TeliaSonera, who launched their service later now has more customers.

In 2007, TPG became the first internet service provider in Australia to launch IPTV. Complementary to its ADSL2+ package this was, and still is free of charge to customers on eligible plans and now offers over 45 local free to air channels and international channels.By 2010, iiNet and Telstra launched IPTV services in conjunction to internet plans but with extra fees.

In 2008, PTCL (Pakistan Telecommunication Company Limited) launched IPTV under the brand name of PTCL Smart TV in Pakistan. This service is available in 150 major cities of the country offering 140 live channels and more than 500 titles for VOD with key features such as:

- EPG (electronic programme guide)
- Parental Control
- Time-Shift Television
- VOD (video on demand)

In 2010, CenturyLink – after acquiring Embarq (2009) and Qwest (2010) – entered five U.S. markets with an IPTV service called Prism. This was after successful test marketing in Florida. During the 2014 Winter Olympics Shortest path bridging (IEEE 802.1aq) was used to deliver 36 IPTV HD Olympic channels.

In 2016, KCTV (Korean Central Television) introduced the Set-top box called "Manbang" (meaning 'everywhere' or 'every direction'), claiming to provide video-on-demand services in North Korea via quasi-internet protocol television (IPTV). With "Manbang", viewers are able to watch five different TV channels in real-time, find information related to the leader's activities and Juche ideology, and read articles from the newspaper Rodong Sinmun and the Korean Central News Agency (KCNA). According to KCTV, viewers can use the service not only in Pyongyang, but also in Sinuiju and Sariwon. Stating that the demands for the equipment are "particularly" high in Sinuiju, with several hundred users in the region.

Promise

The technology was hindered by low broadband penetration and by the relatively high cost of installing wiring capable of transporting IPTV content reliably in the customer's home. However, residential IPTV was expected to grow as broadband was available to more than 200 million households worldwide in 2005.

In December 2009, the FCC began looking into using set-top boxes to make TVs with cable or similar services into network video players. FCC Media Bureau Chief Bill Lake had said earlier that TV and the Internet would soon be the same, but only 75 per cent of homes had computers, while 99 per cent had TV. A 2009 Nielsen survey found 99 per cent of video viewing was done on TV.

Markets

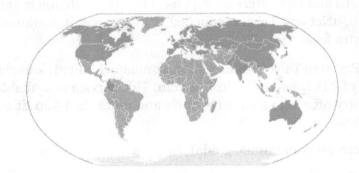

Map of IPTV countries of the world.

Countries where IPTV is available in at least some parts of the country

The number of global IPTV subscribers was expected to grow from 28 million in 2009 to 83 million in 2013. Europe and Asia are the leading territories in terms of the overall number of subscribers. But in terms of service revenues, Europe and North America generate a larger share of global revenue, due to very low average revenue per user (ARPU) in China and India, the fastest growing (and ultimately, the biggest markets) is Asia. The global IPTV market revenues are forecast to grow from US$12 billion in 2009 to US$38 billion in 2013.

Services also launched in Bosnia and Herzegovina, Bulgaria, Pakistan, Canada, Croatia, Lithuania, Moldova, Macedonia, Montenegro, Poland, Mongolia, Romania, Serbia, Slovenia, the Netherlands, Georgia, Greece, Denmark, Finland, Estonia, Czech Republic, Slovakia, Hungary, Norway, Sweden, Iceland, Latvia, Turkey, Colombia,

Chile and Uzbekistan. The United Kingdom launched IPTV early and after a slow initial growth, in February 2009 BT announced that it had reached 398,000 subscribers to its BT Vision service. Claro has launched their own IPTV service called "Claro TV". This service is available in several countries in which they operate, such as Dominican Republic, El Salvador, Guatemala, Honduras, Nicaragua. IPTV is just beginning to grow in Central and Eastern Europe and Latin America, and now it is growing in South Asian countries such as Sri Lanka, Pakistan and especially India. but significant plans exist in countries such as Russia. Kazakhstan introduced its own IPTV services by the national provider Kazakhtelecom JSC and content integrator Alacast under the "iD TV" brand in two major cities Astana and Almaty in 2009 and is about to go nationwide starting 2010.Australian ISP iiNet launched Australia's first IPTV with fetchtv.

The first IPTV service to launch on the Chinese mainland sells under the "BesTV" brand and is currently available in the cities of Shanghai and Harbin. In India, IPTV was launched by Airtel and the government service provider MTNL and BSNL through tie up with AKSH and is available in most of the major cities of the country. Meanwhile, UF Group which is the franchise owner for UFO movies in Southern India plans to offer multiple host of services such as customer's movies on demand, shopping online, video conferencing, media player, e-learning on their single IPTV set top box branded as Emagine.

In Sri Lanka, IPTV was launched by Sri Lanka Telecom (operated by SLT VisionCom) in 2008, under the brand name of PEO TV. This service is available in whole country. The system runs on set-top boxes, smart TV's and various mobile devices thus allowing guests to easily pair their own tablets and smartphones with the TV.

In Pakistan, IPTV was launched by PTCL in 2008, under the brand name of PTCL Smart TV. This service is available in 150 major cities of the country.

In Malaysia, various companies have attempted to launch IPTV services since 2005. Failed PayTV provider MiTV attempted to use an IPTV-over-UHF service but the service failed to take off. Hypp.TV was supposed to use an IPTV-based system, but not true IPTV as it does not provide a set-top box and requires users to view channels using a computer. True IPTV providers available in the country at the moment are Fine TV and DETV. In Q2 2010, Telekom Malaysia launched IPTV services through their fibre to the home product UniFi in select areas. In April 2010, Astro began testing IPTV services on TIME dotCom Berhad's high-speed fibre to the home optical fibre network. In December 2010, Astro began trials with customers in high-rise condominium buildings around the Mont Kiara area. In April 2011, Astro commercially launched its IPTV services under the tag line "The One and Only Line You'll Ever Need", a triple play offering in conjunction with TIME dotCom Berhad that provides all the Astro programming via IPTV, together with voice telephone services and broadband Internet access all through the same fibre optic connection into the customer's home.

In Turkey, TTNET launched IPTV services under the name IPtivibu in 2010. It was available in pilot areas in the cities of Istanbul, İzmir and Ankara. As of 2011, IPTV service is launched as a large-scale commercial service and widely available across the country under the trademark "Tivibu EV". Superonline plans to provide IPTV under the different name "WebTV" in 2011. Türk Telekom started building the fibre optic substructure for IPTV in late 2007.

For Hospitality

Besides targeting the homes, vendors target IPTV services to the hospitality sector. IPTV is a natural progression from the pay-per-view (PPV) and video on demand (VOD) offerings. Some players such as Guest-tek, Locatel, Select-TV, VDA, and Tivus have started offering IPTV to the hotels before moving into the homes. In 2005 GuestTek launched its OneView Media platform providing IPTV to guest rooms in hotels, where users could watch IP-VOD and IPTV from a STB/SBB (Set-back box) connected to the TV. In 2013 Locatel Company launched the most comprehensively integrated IPTV platform available into new markets in Australia, Philippines, Malaysia, Thailand, Ethiopia and Sri Lanka.

Architecture

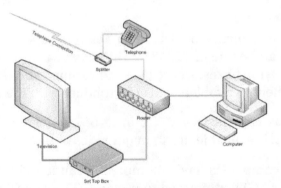

A simplified network diagram for IPTV

Elements

- TV head-end: where live TV channels are encoded, encrypted and delivered in the form of IP multicast streams.

- (VOD) platform: where on-demand video assets are stored and served when a user makes a request in the form of IP unicast stream.

- Interactive portal: allows the user to navigate within the different IPTV services, such as the VOD catalogue.

- Delivery network: the packet switched network that carries IP packets (unicast and multicast).

- Home TV gateway: the piece of equipment at the user's home that terminates the access link from the delivery network.

- User's set-top box: the piece of equipment at the user's home that decodes and decrypts TV and VOD content and displays it on the TV screen.

Architecture of a Video Server Network

Depending on the network architecture of the service provider, there are two main types of video server architecture that can be considered for IPTV deployment: centralised and distributed.

The centralised architecture model is a relatively simple and easy to manage solution. For example, as all contents are stored in centralised servers, it does not require a comprehensive content distribution system. Centralised architecture is generally good for a network that provides relatively small VOD service deployment, has adequate core and edge bandwidth and has an efficient content delivery network (CDN).

Distributed architecture is just as scalable as the centralised model, however it has bandwidth usage advantages and inherent system management features that are essential for managing a larger server network. Operators who plan to deploy a relatively large system should therefore consider implementing a distributed architecture model right from the start. Distributed architecture requires intelligent and sophisticated content distribution technologies to augment effective delivery of multimedia contents over service provider's network.

Home Networks

In many cases, the residential gateway that provides connectivity with the Internet access network is not located close to the IPTV set-top box. This scenario becomes very common as service providers start to offer service packages with multiple set-top boxes per subscriber.

Networking technologies that take advantage of existing home wiring (such as power lines, phone lines or coaxial cables) or of wireless hardware have become common solutions for this problem, although fragmentation in the wired home networking market has limited somewhat the growth in this market.

In December 2008, ITU-T adopted Recommendation G.hn (also known as G.9960), which is a next-generation home networking standard that specifies a common PHY/MAC that can operate over any home wiring (power lines, phone lines or coaxial cables). During 2012 IEC will adopt a prenorm for POF networking at Gigabit speed. This pre standard will specify a PHY that operates at an adaptable bit rate between 100 Mbit/s and 1 Gbit/s depending on the link power budget.

Groups such as the Multimedia over Coax Alliance, HomePlug Powerline Alliance,

Home Phoneline Networking Alliance, and Quasar Alliance (Plastic Optical Fibre) each advocate their own technologies.

IMS Architecture

There is a growing standardisation effort on the use of the 3GPP IP Multimedia Subsystem (IMS) as an architecture for supporting IPTV services in carriers networks. Both ITU-T and ETSI are working on so called "IMS-based IPTV" standards. Carriers will be able to offer both voice and IPTV services over the same core infrastructure and the implementation of services combining conventional TV services with telephony features (e.g. caller ID on the TV screen) will become straightforward. The MultiService Forum recently conducted interoperability of IMS-based IPTV solutions during its GMI event in 2008.

Protocols

IPTV covers both live TV (multicast) as well as stored video-on-demand/VoD (unicast). Playback requires a broadband device connected to either a fixed or wireless IP network in the form of either a standalone personal computer or limited embedded OS device such as a smartphone, touch screen tablet, game console, connected TV or set-top box. Video compression is provided by either a H.263 or H.264 derived codec, audio is compressed via a MDCT based codec and then encapsulated in either an MPEG transport stream or RTP packets or Flash Video packets for live or VoD streaming. IP multicasting allows for live data to be sent to multiple receivers using a single multicast group address. H.264/MPEG-4 AVC is commonly used for internet streaming over higher bit rate standards such as H.261 and H.263 which were more designed for ISDN video conferencing. H.262/MPEG-1/2 is generally not used as the bandwidth required would quite easily saturate a network which is why they are only used in single link broadcast or storage applications.

In standards-based IPTV systems, the primary underlying protocols used are:

- Service provider-based streaming:

  o IGMP for subscribing to a live multicast stream (TV channel) and for changing from one live multicast stream to another (TV channel change). IP multicast operates within LANs (including VLANs) and across WANs also. IP multicast is usually routed in the network core by Protocol Independent Multicast (PIM), setting up correct distribution of multicast streams (TV channels) from their source all the way to the customers who wants to view them, duplicating received packets as needed. On-demand content uses a negotiated unicast connection. Real-time Transport Protocol (RTP) over User Datagram Protocol (UDP) or the lower overhead H.222 transport stream over Transmission Control Protocol (TCP) are generally the preferred methods of encapsulation.

- Web-based unicast only live and VoD streaming:

  o Adobe Flash Player prefers RTMP over TCP with setup and control via either AMF or XML or JSON transactions.

  o Apple iOS uses HLS adaptive bitrate streaming over HTTP with setup and control via an embedded M3U playlist file.

  o Microsoft Silverlight uses smooth streaming (adaptive bitrate streaming) over HTTP.

- Web-based multicast live and unicast VoD streaming:

  o The Internet Engineering Task Force (IETF) recommends RTP over UDP or TCP transports with setup and control using RTSP over TCP.

- Connected TVs, game consoles, set-top boxes and network personal video recorders:

  o local network content uses UPnP AV for unicast via HTTP over TCP or for multicast live RTP over UDP.

  o Web-based content is provided through either inline Web plug-ins or a television broadcast-based application that uses a middleware language such as MHEG-5 that triggers an event such as loading an inline Web browser using an Adobe Flash Player plug-in.

A telecommunications company IPTV service is usually delivered over an investment-heavy walled garden network.

Local IPTV, as used by businesses for audio visual AV distribution on their company networks is typically based on a mixture of:

1. Conventional TV reception equipment and IPTV encoders

2. IPTV gateways that take broadcast MPEG channels and IP wrap them to create multicast streams.

Via Satellite

Although IPTV and conventional satellite TV distribution have been seen as complementary technologies, they are likely to be increasingly used together in hybrid IPTV networks that deliver the highest levels of performance and reliability. IPTV is largely neutral to the transmission medium, and IP traffic is already routinely carried by satellite for Internet backbone trunking and corporate VSAT networks. The use of satellite to carry IP is fundamental to overcoming the greatest shortcoming of IPTV over terrestrial cables – the speed/bandwidth of the connection, as well as availability.

The copper twisted pair cabling that forms the last mile of the telephone and broadband network in many countries is not able to provide a sizeable proportion of the population with an IPTV service that matches even existing terrestrial or satellite digital TV distribution. For a competitive multi-channel TV service, a connection speed of 20 Mbit/s is likely to be required, but unavailable to most potential customers. The increasing popularity of high-definition television (with twice the data rate of SD video) increases connection speed requirements, or limits IPTV service quality and connection eligibility even further.

However, satellites are capable of delivering in excess of 100 Gbit/s via multi-spot beam technologies, making satellite a clear emerging technology for implementing IPTV networks. Satellite distribution can be included in an IPTV network architecture in several ways. The simplest to implement is an IPTV-direct to home (DTH) architecture, in which hybrid DVB-broadband set-top boxes in subscriber homes integrate satellite and IP reception to give near-infinite bandwidth with return channel capabilities. In such a system, many live TV channels may be multicast via satellite (IP-encapsulated or as conventional DVB digital TV) with stored video-on-demand transmission via the broadband connection. Arqiva's Satellite Media Solutions Division suggests "IPTV works best in a hybrid format. For example, you would use broadband to receive some content and satellite to receive other, such as live channels".

Hybrid IPTV

Hybrid IPTV refers to the combination of traditional broadcast TV services and video delivered over either managed IP networks or the public Internet. It is an increasing trend in both the consumer and pay TV [operator] markets.

Hybrid IPTV has grown in popularity in recent years as a result of two major drivers. Since the emergence of online video aggregation sites, like YouTube and Vimeo in the mid-2000s, traditional pay TV operators have come under increasing pressure to provide their subscribers with a means of viewing Internet-based video [both professional and user-generated] on their televisions. At the same time, specialist IP-based operators [often telecommunications providers] have looked for ways to offer analogue and digital terrestrial services to their operations, without adding either additional cost or complexity to their transmission operations. Bandwidth is a valuable asset for operators, so many have looked for alternative ways to deliver these new services without investing in additional network infrastructures.

A hybrid set-top allows content from a range of sources, including terrestrial broadcast, satellite, and cable, to be brought together with video delivered over the Internet via an Ethernet connection on the device. This enables television viewers to access a greater variety of content on their TV sets, without the need for a separate box for each service.

Hybrid IPTV set-top boxes also enable users to access a range of advanced interactive services, such as VOD / catch-up TV, as well as Internet applications, including video telephony, surveillance, gaming, shopping, e-government accessed via a television set.

From a pay-TV operator's perspective, a hybrid IPTV set-top box gives them greater long-term flexibility by enabling them to deploy new services and applications as and when consumers require, most often without the need to upgrade equipment or for a technician to visit and reconfigure or swap out the device. This reduces the cost of launching new services, increases speed to market and limits disruption for consumers.

The Hybrid Broadcast Broadband TV (HbbTV) consortium of industry companies is currently promoting and establishing an open European standard for hybrid set-top boxes for the reception of broadcast and broadband digital TV and multimedia applications with a single user interface. These trends led to the development of Hybrid Broadcast Broadband TV set-top boxes that included both a broadcast tuner and an Internet connection – usually an Ethernet port. The first commercially available hybrid IPTV set-top box was developed by Advanced Digital Broadcast, a developer of digital television hardware and software, in 2005. The platform was developed for Spanish pay TV operator Telefonica, and used as part of its Movistar TV service, launched to subscribers at the end of 2005.

An alternative approach is the IPTV version of the Headend in the Sky cable TV solution. Here, multiple TV channels are distributed via satellite to the ISP or IPTV provider's point of presence (POP) for IP-encapsulated distribution to individual subscribers as required by each subscriber.

This can provide a huge selection of channels to subscribers without overburdening Internet trunking to the POP, and enables an IPTV service to be offered to small or remote operators outside the reach of terrestrial high speed broadband connection. An example is a network combining fibre and satellite distribution via an SES New Skies satellite of 95 channels to Latin America and the Caribbean, operated by IPTV Americas.

While the future development of IPTV probably lies with a number of coexisting architectures and implementations, it is clear that broadcasting of high bandwidth applications such as IPTV is accomplished more efficiently and cost-effectively using satellite and it is predicted that the majority of global IPTV growth will be fuelled by hybrid networks.

Advantages

The Internet protocol-based platform offers significant advantages, including the ability to integrate television with other IP-based services like high speed Internet access and VoIP.

A switched IP network also allows for the delivery of significantly more content and functionality. In a typical TV or satellite network, using broadcast video technology, all the content constantly flows downstream to each customer, and the customer switches the content at the set-top box. The customer can select from as many choices as the telecomms, cable or satellite company can stuff into the "pipe" flowing into the home. A switched IP network works differently. Content remains in the network, and only the content the customer selects is sent into the customer's home. That frees up bandwidth, and the customer's choice is less restricted by the size of the "pipe" into the home. This also implies that the customer's privacy could be compromised to a greater extent than is possible with traditional TV or satellite networks. It may also provide a means to hack into, or at least disrupt the private network.

Economics

The cable industry's expenditures of approximately $1 billion per year are based on network updates to accommodate higher data speeds. Most operators use 2–3 channels to support maximum data speeds of 50 Mbit/s to 100 Mbit/s. However, because video streams require a high bit rate for much longer periods of time, the expenditures to support high amounts of video traffic will be much greater. This phenomenon is called persistency. Data persistency is routinely 5% while video persistency can easily reach 50%. As video traffic continues to grow, this means that significantly more CMTS downstream channels will be required to carry this video content. Based on today's market, it is likely that industry expenditures for CMTS expansion could exceed $2 billion a year, virtually all of that expenditure being driven by video traffic. Adoption of IPTV for carrying the majority of this traffic could save the industry approximately 75% of this capital expenditure.

Interactivity

An IP-based platform also allows significant opportunities to make the TV viewing experience more interactive and personalised. The supplier may, for example, include an interactive programme guide that allows viewers to search for content by title or actor's name, or a picture-in-picture functionality that allows them to "channel surf" without leaving the programme they're watching. Viewers may be able to look up a player's stats while watching a sports game, or control the camera angle. They also may be able to access photos or music from their PC on their television, use a wireless phone to schedule a recording of their favourite show, or even adjust parental controls so their child can watch a documentary for a school report, while they're away from home.

In order that there can take place an interaction between the receiver and the transmitter, a feedback channel is needed. Due to this, terrestrial, satellite, and cable networks for television do not allow interactivity. However, interactivity with those networks can be possible by combining TV networks with data networks such as the Internet or a mobile communication network.

Video-on-demand

IPTV technology is bringing video on demand (VoD) to television, which permits a customer to browse an online programme or film catalogue, to watch trailers and to then select a selected recording. The playout of the selected item starts nearly instantaneously on the customer's TV or PC.

Technically, when the customer selects the movie, a point-to-point unicast connection is set up between the customer's decoder (set-top box or PC) and the delivering streaming server. The signalling for the trick play functionality (pause, slow-motion, wind/rewind etc.) is assured by RTSP (Real Time Streaming Protocol).

The most common codecs used for VoD are MPEG-2, MPEG-4 and VC-1.

In an attempt to avoid content piracy, the VoD content is usually encrypted. Whilst encryption of satellite and cable TV broadcasts is an old practice, with IPTV technology it can effectively be thought of as a form of Digital rights management. A film that is chosen, for example, may be playable for 24 hours following payment, after which time it becomes unavailable.

IPTV-based Converged Services

Another advantage is the opportunity for integration and convergence. This opportunity is amplified when using IMS-based solutions. Converged services implies interaction of existing services in a seamless manner to create new value added services. One example is on-screen Caller ID, getting Caller ID on a TV and the ability to handle it (send it to voice mail, etc.). IP-based services will help to enable efforts to provide consumers anytime-anywhere access to content over their televisions, PCs and cell phones, and to integrate services and content to tie them together. Within businesses and institutions, IPTV eliminates the need to run a parallel infrastructure to deliver live and stored video services.

Limitations

IPTV is sensitive to packet loss and delays if the streamed data is unreliable. IPTV has strict minimum speed requirements in order to facilitate the right number of frames per second to deliver moving pictures. This means that the limited connection speed and bandwidth available for a large IPTV customer base can reduce the service quality delivered.

Although a few countries have very high-speed broadband-enabled populations, such as South Korea with 6 million homes benefiting from a minimum connection speed of 100 Mbit/s, in other countries (such as the UK) legacy networks struggle to provide 3–5 Mbit/s and so simultaneous provision to the home of TV channels, VOIP and Internet access may not be viable. The last-mile delivery for IPTV usually has a bandwidth restriction that only allows a small number of simultaneous TV channel streams – typically from one to three – to be delivered.

Streaming IPTV across wireless links within the home has proved troublesome; not due to bandwidth limitations as many assume, but due to issues with multipath and reflections of the RF signal carrying the IP data packets. An IPTV stream is sensitive to packets arriving at the right time and in the right order. Improvements in wireless technology are now starting to provide equipment to solve the problem.

Due to the limitations of wireless, most IPTV service providers today use wired home networking technologies instead of wireless technologies like IEEE 802.11. Service providers such as AT&T (which makes extensive use of wireline home networking as part of its AT&T U-verse IPTV service) have expressed support for the work done in this direction by ITU-T, which has adopted Recommendation G.hn (also known as G.9960), which is a next-generation home networking standard that specifies a common PHY/MAC that can operate over any home wiring (power lines, phone lines or coaxial cables).

Latency

The latency inherent in the use of satellite Internet is often held up as reason why satellites cannot be successfully used for IPTV. In practice, however, latency is not an important factor for IPTV, since it is a service that does not require real-time transmission, as is the case with telephony or videoconferencing services.

It is the latency of response to requests to change channel, display an EPG, etc. that most affects customers' perceived quality of service, and these problems affect satellite IPTV no more than terrestrial IPTV. Command latency problems, faced by terrestrial IPTV networks with insufficient bandwidth as their customer base grows, may be solved by the high capacity of satellite distribution.

Satellite distribution does suffer from latency – the time for the signal to travel up from the hub to the satellite and back down to the user is around 0.25 seconds, and cannot be reduced. However, the effects of this delay are mitigated in real-life systems using data compression, TCP-acceleration, and HTTP pre-fetching.

Satellite latency can be detrimental to especially time-sensitive applications such as on-line gaming (although it only seriously affects the likes of first-person shooters while many MMOGs can operate well over satellite Internet), but IPTV is typically a simplex operation (one-way transmission) and latency is not a critical factor for video transmission.

Existing video transmission systems of both analogue and digital formats already introduce known quantifiable delays. Existing DVB TV channels that simulcast by both terrestrial and satellite transmissions experience the same 0.25-second delay difference between the two services with no detrimental effect, and it goes unnoticed by viewers.

Bandwidth Requirements

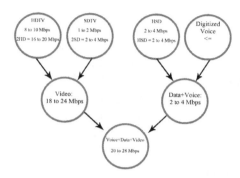

Bandwidth capacity for simultaneously two HDTV streams, two SD streams, additional to HSD and voice

Digital video is a combination of sequence of digital images, and they are made up of pixels or picture elements. Each pixel has two values, which are luminance and chrominance. Luminance is representing intensity of the pixel; chrominance represents the colour of the pixel. Three bytes would be used to represent the colour of the high quality image for a true colour technique. A sequence of images is creating the digital video, in that case, images are called as frames.

Movies use 24 frames per second; however, the rate of the frames can change according to territories' electrical systems so that there are different kinds of frame rates, for instance, North America is using approximately 30 frames per second where the Europe television frame rate is 25 frames per second. Each digital video has dimensions width and height; when referred to analogue television, the dimension for SDTV is 720×480 pixels, on the other hand, numerous HDTV requires 1920×1080 pixels. Moreover, whilst for SDTV, two bytes (16 bits) is enough to create the colour depth, HDTV requires three bytes (24 bits) to create the colour depth.

Thereby, with a rate of 30 frames/second, the uncompressed data rate for SDTV becomes 30×640×480×16, in other words, 147,456,000 bits per second. Moreover, for HDTV, at the same frame rate, uncompressed date rate becomes 30×1920×1080×24 or 1,492,992,000 bits per second. With that simple calculation, it is obvious that without using a lossy compression methods service provider's service delivery to the subscribers is limited.

There is no absolute answer for the bandwidth requirement for the IPTV service because the bandwidth requirement is increasing due to the devices inside the household. Thus, currently compressed HDTV content can be delivered at a data rate between 8 and 10 Mbit/s, but if the home of the consumer equipped with several HDTV outputs, this rate will be multiplied respectively.

The high-speed data transfer will increase the needed bandwidth for the viewer, at least 2 Mbit/s is needed to use web-based applications on the computer. Additionally to that,

64 kbit/s is required to use landline telephone for the property. In minimal usage, to receive an IPTV triple-play service requires 13 Mbit/s to process in a household.

Privacy Implications

Due to limitations in bandwidth, an IPTV channel is delivered to the user one at a time, as opposed to the traditional multiplexed delivery. Changing a channel requires requesting the head-end server to provide a different broadcast stream, much like VOD (For VOD the stream is delivered using unicast, for the normal TV signal multicast is used). This could enable the service provider to accurately track each and every programme watched and the duration of watching for each viewer; broadcasters and advertisers could then understand their audience and programming better with accurate data and targeted advertising.

In conjunction with regulatory differences between IPTV and cable TV, this tracking could pose a threat to privacy according to critics. For IP multicast scenarios, since a particular multicast group (TV channel) needs to be requested before it can be viewed, the same privacy concerns apply.

Vendors

A small number of companies supply most current IPTV systems. Some, such as Movistar TV, were formed by telecoms operators themselves, to minimise external costs, a tactic also used by PCCW of Hong Kong. Some major telecoms vendors are also active in this space, notably Alcatel-Lucent (sometimes working with Movistar TV), Ericsson (notably since acquiring Tandberg Television), NEC, Accenture (Accenture Video Solution), Thomson, Huawei, and ZTE, as are some IT houses, led by Microsoft. California-based UTStarcom, Inc., Tennessee-based Worley Consulting, Tokyo-based The New Media Group, Malaysian-based Select-TV and Oslo/Norway-based SnapTV also offer end-to-end networking infrastructure for IPTV-based services, and Hong Kong-based BNS Ltd. provides turnkey open platform IPTV technology solutions. Global sales of IPTV systems exceeded 2 billion USD in 2007.

Hospitality IPTV Ltd, having established many closed network IPTV systems, expanded in 2013 to OTT delivery platforms for markets in New Zealand, Australia and Asia Pacific region.

Google Fiber offers an IPTV service in various US cities which includes up to 1 Gigabit-speed internet and over 290 channels depending of package via the fibre optic network being built out in KCK and KCMO.

Many of these IPTV solution vendors participated in the biennial Global MSF Interoperability 2008 (GMI) event which was coordinated by the MultiService Forum (MSF) at five sites worldwide from 20 to 31 October 2008. Test equipment vendors including Netrounds, Codenomicon, Empirix, Ixia, Mu Dynamics and Spirent joined solution vendors such as the companies listed above in one of the largest IPTV proving grounds ever deployed.

Service Bundling

For residential users, IPTV is often provided in conjunction with video on demand and may be bundled with Internet services such as Internet access and Voice over Internet Protocol (VoIP) telecommunications services. Commercial bundling of IPTV, VoIP and Internet access is sometimes referred to in marketing as *triple play* service. When these three are offered with cellular service, the combined service may be referred to as *quadruple play*.

Regulation

Historically, broadcast television has been regulated differently from telecommunications. As IPTV allows TV and VoD to be transmitted over IP networks, new regulatory issues arise. Professor Eli M. Noam highlights in his report "TV or Not TV: Three Screens, One Regulation?" some of the key challenges with sector specific regulation that is becoming obsolete due to convergence in this field.

Internet Access

Internet access is the process that enables individuals and organisations to connect to the Internet using computer terminals, computers, mobile devices, sometimes via computer networks. Once connected to the Internet, users can access Internet services, such as email and the World Wide Web. Internet service providers (ISPs) offer Internet access through various technologies that offer a wide range of data signaling rates (speeds).

Consumer use of the Internet first became popular through dial-up Internet access in the 1990s. By the first decade of the 21st century, many consumers in developed nations used faster, broadband Internet access technologies. By 2014 this was almost ubiquitous worldwide, with a global average connection speed exceeding 4 Mbit/s.

History

The Internet developed from the ARPANET, which was funded by the US government to support projects within the government and at universities and research laboratories in the US – but grew over time to include most of the world's large universities and the research arms of many technology companies. Use by a wider audience only came in 1995 when restrictions on the use of the Internet to carry commercial traffic were lifted.

In the early to mid-1980s, most Internet access was from personal computers and workstations directly connected to local area networks or from dial-up connections using modems and analog telephone lines. LANs typically operated at 10 Mbit/s, while modem data-rates grew from 1200 bit/s in the early 1980s, to 56 kbit/s by the late 1990s. Initially, dial-up connections were made from terminals or computers running terminal emulation software to terminal servers on LANs. These dial-up connections did not support end-to-end use of the Internet protocols and only provided terminal to host connections. The introduction of network access servers supporting the Serial Line Internet Protocol (SLIP) and later the point-to-point protocol (PPP) extended the Internet protocols and made the full range of Internet services available to dial-up users; although slower, due to the lower data rates available using dial-up.

Broadband Internet access, often shortened to just broadband, is simply defined as "Internet access that is always on, and faster than the traditional dial-up access" and so covers a wide range of technologies. Broadband connections are typically made using a computer's built in Ethernet networking capabilities, or by using a NIC expansion card.

Most broadband services provide a continuous "always on" connection; there is no dial-in process required, and it does not interfere with voice use of phone lines. Broadband provides improved access to Internet services such as:

- Faster world wide web browsing

- Faster downloading of documents, photographs, videos, and other large files

- Telephony, radio, television, and videoconferencing

- Virtual private networks and remote system administration

- Online gaming, especially massively multiplayer online role-playing games which are interaction-intensive

In the 1990s, the National Information Infrastructure initiative in the U.S. made broadband Internet access a public policy issue. In 2000, most Internet access to homes was provided using dial-up, while many businesses and schools were using broadband connections. In 2000 there were just under 150 million dial-up subscriptions in the 34 OECD countries and fewer than 20 million broadband subscriptions. By 2004, broadband had grown and dial-up had declined so that the number of subscriptions were roughly equal at 130 million each. In 2010, in the OECD countries, over 90% of the Internet access subscriptions used broadband, broadband had grown to more than 300 million subscriptions, and dial-up subscriptions had declined to fewer than 30 million.

The broadband technologies in widest use are ADSL and cable Internet access. Newer technologies include VDSL and optical fibre extended closer to the subscriber in both telephone and cable plants. Fibre-optic communication, while only recently being used in premises and to the curb schemes, has played a crucial role in enabling broadband Internet access by making transmission of information at very high data rates over longer distances much more cost-effective than copper wire technology.

In areas not served by ADSL or cable, some community organizations and local governments are installing Wi-Fi networks. Wireless and satellite Internet are often used in rural, undeveloped, or other hard to serve areas where wired Internet is not readily available.

Newer technologies being deployed for fixed (stationary) and mobile broadband access include WiMAX, LTE, and fixed wireless, e.g., Motorola Canopy.

Starting in roughly 2006, mobile broadband access is increasingly available at the consumer level using "3G" and "4G" technologies such as HSPA, EV-DO, HSPA+, and LTE.

Availability

In addition to access from home, school, and the workplace Internet access may be available from public places such as libraries and Internet cafes, where computers with Internet connections are available. Some libraries provide stations for physically connecting users' laptops to local area networks (LANs).

Wireless Internet access points are available in public places such as airport halls, in some cases just for brief use while standing. Some access points may also provide coin operated computers. Various terms are used, such as "public Internet kiosk", "public access terminal", and "Web payphone". Many hotels also have public terminals, usually fee based.

Coffee shops, shopping malls, and other venues increasingly offer wireless access to computer networks, referred to as hotspots, for users who bring their own wireless-enabled devices such as a laptop or PDA. These services may be free to all, free to customers only, or fee-based. A Wi-Fi hotspot need not be limited to a confined location since multiple ones combined can cover a whole campus or park, or even an entire city can be enabled.

Additionally, Mobile broadband access allows smart phones and other digital devices to connect to the Internet from any location from which a mobile phone call can be made, subject to the capabilities of that mobile network.

Speed

The bit rates for dial-up modems range from as little as 110 bit/s in the late 1950s, to a maximum of from 33 to 64 kbit/s (V.90 and V.92) in the late 1990s. Dial-up connections generally require the dedicated use of a telephone line. Data compression can boost the effective bit rate for a dial-up modem connection to from 220 (V.42bis) to 320 (V.44) kbit/s. However, the effectiveness of data compression is quite variable, depending on the type of data being sent, the condition of the telephone line, and a number of other factors. In reality, the overall data rate rarely exceeds 150 kbit/s.

Broadband technologies supply considerably higher bit rates than dial-up, generally without disrupting regular telephone use. Various minimum data rates and maximum latencies have been used in definitions of broadband, ranging from 64 kbit/s up to 4.0 Mbit/s. In 1988 the CCITT standards body defined "broadband service" as requiring transmission channels capable of supporting bit rates greater than the primary rate which ranged from about 1.5 to 2 Mbit/s. A 2006 Organization for Economic Co-operation and Development (OECD) report defined broadband as having download data transfer rates equal to or faster than 256 kbit/s. And in 2015 the U.S. Federal Communications Commission (FCC) defined "Basic Broadband" as data transmission speeds of at least 25 Mbit/s downstream (from the Internet to the user's computer) and 3 Mbit/s upstream (from the user's computer to the Internet). The trend is to raise the threshold of the broadband definition as higher data rate services become available.

The higher data rate dial-up modems and many broadband services are "asymmetric"—supporting much higher data rates for download (toward the user) than for upload (toward the Internet).

Data rates, including those given in this article, are usually defined and advertised in terms of the maximum or peak download rate. In practice, these maximum data rates are not always reliably available to the customer. Actual end-to-end data rates can be lower due to a number of factors. In late June 2016, internet connection speeds averaged about 6 Mbit/s globally. Physical link quality can vary with distance and for wireless access with terrain, weather, building construction, antenna placement, and interference from other radio sources. Network bottlenecks may exist at points anywhere on the path from the end-user to the remote server or service being used and not just on the first or last link providing Internet access to the end-user.

Network Congestion

Users may share access over a common network infrastructure. Since most users do not use their full connection capacity all of the time, this aggregation strategy (known as contended service) usually works well and users can burst to their full data rate at least for brief periods. However, peer-to-peer (P2P) file sharing and high-quality streaming video can require high data-rates for extended periods, which violates these assumptions and can cause a service to become oversubscribed, resulting in congestion and poor performance. The TCP protocol includes flow-control mechanisms that automatically throttle back on the bandwidth being used during periods of network congestion. This is fair in the sense that all users that experience congestion receive less bandwidth, but it can be frustrating for customers and a major problem for ISPs. In some cases the amount of bandwidth actually available may fall below the threshold required to support a particular service such as video conferencing or streaming live video—effectively making the service unavailable.

When traffic is particularly heavy, an ISP can deliberately throttle back the bandwidth available to classes of users or for particular services. This is known as traffic shaping and careful use can ensure a better quality of service for time critical services even on extremely busy networks. However, overuse can lead to concerns about fairness and network neutrality or even charges of censorship, when some types of traffic are severely or completely blocked.

Outages

An Internet blackout or outage can be caused by local signaling interruptions. Disruptions of submarine communications cables may cause blackouts or slowdowns to large areas, such as in the 2008 submarine cable disruption. Less-developed countries are more vulnerable due to a small number of high-capacity links. Land cables are also vulnerable, as in 2011 when a woman digging for scrap metal severed most connectivity for the nation of Armenia. Internet blackouts affecting almost entire countries can be achieved by governments as a form of Internet censorship, as in the blockage of the Internet in Egypt, whereby approximately 93% of networks were without access in 2011 in an attempt to stop mobilization for anti-government protests.

On April 25, 1997, due to a combination of human error and software bug, an incorrect routing table at MAI Network Service (a Virginia Internet Service Provider) propagated across backbone routers and caused major disruption to Internet traffic for a few hours.

Technologies

When the Internet is accessed using a modem, digital data is converted to analog for transmission over analog networks such as the telephone and cable networks. A computer or other device accessing the Internet would either be connected directly to a modem that communicates with an Internet service provider (ISP) or the modem's Internet connection would be shared via a Local Area Network (LAN) which provides access in a limited area such as a home, school, computer laboratory, or office building.

Although a connection to a LAN may provide very high data-rates within the LAN, actual Internet access speed is limited by the upstream link to the ISP. LANs may be wired or wireless. Ethernet over twisted pair cabling and Wi-Fi are the two most common technologies used to build LANs today, but ARCNET, Token Ring, Localtalk, FDDI, and other technologies were used in the past.

Ethernet is the name of the IEEE 802.3 standard for physical LAN communication and Wi-Fi is a trade name for a wireless local area network (WLAN) that uses one of the IEEE 802.11 standards. Ethernet cables are interconnected via switches & routers. Wi-Fi networks are built using one or more wireless antenna called access points.

Many "modems" provide the additional functionality to host a LAN so most Internet access today is through a LAN, often a very small LAN with just one or two devices attached. And while LANs are an important form of Internet access, this raises the question of how and at what data rate the LAN itself is connected to the rest of the global Internet. The technologies described below are used to make these connections.

Hardwired Broadband Access

The term broadband includes a broad range of technologies, all of which provide higher data rate access to the Internet. The following technologies use wires or cables in contrast to wireless broadband described later.

Dial-up Access

Dial-up Internet access uses a modem and a phone call placed over the public switched telephone network (PSTN) to connect to a pool of modems operated by an ISP. The modem converts a computer's digital signal into an analog signal that travels over a phone line's local loop until it reaches a telephone company's switching facilities or central office (CO) where it is switched to another phone line that connects to another modem at the remote end of the connection.

Operating on a single channel, a dial-up connection monopolizes the phone line and is one of the slowest methods of accessing the Internet. Dial-up is often the only form of Internet access available in rural areas as it requires no new infrastructure beyond the already existing telephone network, to connect to the Internet. Typically, dial-up connections do not exceed a speed of 56 kbit/s, as they are primarily made using modems that operate at a maximum data rate of 56 kbit/s downstream (towards the end user) and 34 or 48 kbit/s upstream (toward the global Internet).

Multilink Dial-up

Multilink dial-up provides increased bandwidth by channel bonding multiple dial-up connections and accessing them as a single data channel. It requires two or more modems, phone lines, and dial-up accounts, as well as an ISP that supports multilinking – and of course any line and data charges are also doubled. This inverse multiplexing option was briefly popular with some high-end users before ISDN, DSL and other technologies became available. Diamond and other vendors created special modems to support multilinking.

Integrated Services Digital Network

Integrated Services Digital Network (ISDN) is a switched telephone service capable of transporting voice and digital data, is one of the oldest Internet access methods. ISDN has been used for voice, video conferencing, and broadband data applications. ISDN was very popular in Europe, but less common in North America. Its use peaked in the late 1990s before the availability of DSL and cable modem technologies.

Basic rate ISDN, known as ISDN-BRI, has two 64 kbit/s "bearer" or "B" channels. These channels can be used separately for voice or data calls or bonded together to provide a 128 kbit/s service. Multiple ISDN-BRI lines can be bonded together to provide data rates above 128 kbit/s. Primary rate ISDN, known as ISDN-PRI, has 23 bearer channels (64 kbit/s each) for a combined data rate of 1.5 Mbit/s (US standard). An ISDN E1 (European standard) line has 30 bearer channels and a combined data rate of 1.9 Mbit/s.

Leased Lines

Leased lines are dedicated lines used primarily by ISPs, business, and other large enterprises to connect LANs and campus networks to the Internet using the existing infrastructure of the public telephone network or other providers. Delivered using wire, optical fiber, and radio, leased lines are used to provide Internet access directly as well as the building blocks from which several other forms of Internet access are created.

T-carrier technology dates to 1957 and provides data rates that range from 56 and 64 kbit/s (DS0) to 1.5 Mbit/s (DS1 or T1), to 45 Mbit/s (DS3 or T3). A T1 line carries 24 voice or data channels (24 DS0s), so customers may use some channels for data and others for voice traffic or use all 24 channels for clear channel data. A DS3 (T3) line carries 28 DS1 (T1) channels. Fractional T1 lines are also available in multiples of a DS0 to provide data rates between 56 and 1,500 kbit/s. T-carrier lines require special termination equipment that may be separate from or integrated into a router or switch and which may be purchased or leased from an ISP. In Japan the equivalent standard is J1/J3. In Europe, a slightly different standard, E-carrier, provides 32 user channels (64 kbit/s) on an E1 (2.0 Mbit/s) and 512 user channels or 16 E1s on an E3 (34.4 Mbit/s).

Synchronous Optical Networking (SONET, in the U.S. and Canada) and Synchronous Digital Hierarchy (SDH, in the rest of the world) are the standard multiplexing protocols used to carry high-data-rate digital bit-streams over optical fiber using lasers or highly coherent light from light-emitting diodes (LEDs). At lower transmission rates data can also be transferred via an electrical interface. The basic unit of framing is an OC-3c (optical) or STS-3c (electrical) which carries 155.520 Mbit/s. Thus an OC-3c will carry three OC-1 (51.84 Mbit/s) payloads each of which has enough capacity to include a full DS3. Higher data rates are delivered in OC-3c multiples of four providing OC-12c (622.080 Mbit/s), OC-48c (2.488 Gbit/s), OC-192c (9.953 Gbit/s), and OC-768c (39.813 Gbit/s). The "c" at the end of the OC labels stands for "concatenated" and indicates a single data stream rather than several multiplexed data streams.

The 1, 10, 40, and 100 gigabit Ethernet (GbE, 10 GbE, 40/100 GbE) IEEE standards (802.3) allow digital data to be delivered over copper wiring at distances to 100 m and over optical fiber at distances to 40 km.

Cable Internet Access

Cable Internet provides access using a cable modem on hybrid fiber coaxial wiring orig-inally developed to carry television signals. Either fiber-optic or coaxial copper cable may connect a node to a customer's location at a connection known as a cable drop. In a cable modem termination system, all nodes for cable subscribers in a neighborhood connect to a cable company's central office, known as the "head end." The cable com-pany then connects to the Internet using a variety of means – usually fiber optic cable or digital satellite and microwave transmissions. Like DSL, broadband cable provides a continuous connection with an ISP.

Downstream, the direction toward the user, bit rates can be as much as 400 Mbit/s for business connections, and 250 Mbit/s for residential service in some countries. Up-stream traffic, originating at the user, ranges from 384 kbit/s to more than 20 Mbit/s. Broadband cable access tends to service fewer business customers because existing tele-vision cable networks tend to service residential buildings and commercial buildings do not always include wiring for coaxial cable networks. In addition, because broad-band cable subscribers share the same local line, communications may be intercepted by neighboring subscribers. Cable networks regularly provide encryption schemes for data traveling to and from customers, but these schemes may be thwarted.

Digital Subscriber Line (DSL, ADSL, SDSL, and VDSL)

Digital Subscriber Line (DSL) service provides a connection to the Internet through the telephone network. Unlike dial-up, DSL can operate using a single phone line without preventing normal use of the telephone line for voice phone calls. DSL uses the high frequencies, while the low (audible) frequencies of the line are left free for regular tele-phone communication. These frequency bands are subsequently separated by filters installed at the customer's premises.

DSL originally stood for "digital subscriber loop". In telecommunications marketing, the term digital subscriber line is widely understood to mean Asymmetric Digital Sub-scriber Line (ADSL), the most commonly installed variety of DSL. The data throughput of consumer DSL services typically ranges from 256 kbit/s to 20 Mbit/s in the direction to the customer (downstream), depending on DSL technology, line conditions, and ser-vice-level implementation. In ADSL, the data throughput in the upstream direction, (i.e. in the direction to the service provider) is lower than that in the downstream di-rection (i.e. to the customer), hence the designation of asymmetric. With a symmetric digital subscriber line (SDSL), the downstream and upstream data rates are equal.

Very-high-bit-rate digital subscriber line (VDSL or VHDSL, ITU G.993.1) is a digi-tal subscriber line (DSL) standard approved in 2001 that provides data rates up to 52 Mbit/s downstream and 16 Mbit/s upstream over copper wires and up to 85 Mbit/s down- and upstream on coaxial cable. VDSL is capable of supporting applications such as high-definition television, as well as telephone services (voice over IP) and general Internet access, over a single physical connection.

VDSL2 (ITU-T G.993.2) is a second-generation version and an enhancement of VDSL. Approved in February 2006, it is able to provide data rates exceeding 100 Mbit/s simultaneously in both the upstream and downstream directions. However, the maximum data rate is achieved at a range of about 300 meters and performance degrades as distance and loop attenuation increases.

DSL Rings

DSL Rings (DSLR) or Bonded DSL Rings is a ring topology that uses DSL technology over existing copper telephone wires to provide data rates of up to 400 Mbit/s.

Fiber to the Home

Fiber-to-the-home (FTTH) is one member of the Fiber-to-the-x (FTTx) family that includes Fiber-to-the-building or basement (FTTB), Fiber-to-the-premises (FTTP), Fiber-to-the-desk (FTTD), Fiber-to-the-curb (FTTC), and Fiber-to-the-node (FTTN). These methods all bring data closer to the end user on optical fibers. The differences between the methods have mostly to do with just how close to the end user the delivery on fiber comes. All of these delivery methods are similar to hybrid fiber-coaxial (HFC) systems used to provide cable Internet access.

The use of optical fiber offers much higher data rates over relatively longer distances. Most high-capacity Internet and cable television backbones already use fiber optic technology, with data switched to other technologies (DSL, cable, POTS) for final delivery to customers.

Australia began rolling out its National Broadband Network across the country using fiber-optic cables to 93 percent of Australian homes, schools, and businesses. The project was abandoned by the subsequent LNP government, in favour of a hybrid FTTN design, which turned out to be more expensive and introduced delays. Similar efforts are underway in Italy, Canada, India, and many other countries.

Power-line Internet

Power-line Internet, also known as Broadband over power lines (BPL), carries Internet data on a conductor that is also used for electric power transmission. Because of the extensive power line infrastructure already in place, this technology can provide people in rural and low population areas access to the Internet with little cost in terms of new transmission equipment, cables, or wires. Data rates are asymmetric and generally range from 256 kbit/s to 2.7 Mbit/s.

Because these systems use parts of the radio spectrum allocated to other over-the-air communication services, interference between the services is a limiting factor in the introduction of power-line Internet systems. The IEEE P1901 standard specifies that all power-line protocols must detect existing usage and avoid interfering with it.

Power-line Internet has developed faster in Europe than in the U.S. due to a historical difference in power system design philosophies. Data signals cannot pass through the step-down transformers used and so a repeater must be installed on each transformer. In the U.S. a transformer serves a small cluster of from one to a few houses. In Europe, it is more common for a somewhat larger transformer to service larger clusters of from 10 to 100 houses. Thus a typical U.S. city requires an order of magnitude more repeaters than in a comparable European city.

ATM and Frame Relay

Asynchronous Transfer Mode (ATM) and Frame Relay are wide-area networking standards that can be used to provide Internet access directly or as building blocks of other access technologies. For example, many DSL implementations use an ATM layer over the low-level bitstream layer to enable a number of different technologies over the same link. Customer LANs are typically connected to an ATM switch or a Frame Relay node using leased lines at a wide range of data rates.

While still widely used, with the advent of Ethernet over optical fiber, MPLS, VPNs and broadband services such as cable modem and DSL, ATM and Frame Relay no longer play the prominent role they once did.

Wireless Broadband Access

Wireless broadband is used to provide both fixed and mobile Internet access with the following technologies.

Satellite Broadband

Satellite Internet access via VSAT in Ghana

Satellite Internet access provides fixed, portable, and mobile Internet access. Data rates range from 2 kbit/s to 1 Gbit/s downstream and from 2 kbit/s to 10 Mbit/s upstream. In the northern hemisphere, satellite antenna dishes require a clear line of sight to the southern sky, due to the equatorial position of all geostationary satellites. In the southern hemisphere, this situation is reversed, and dishes are pointed north. Service can be adversely affected by moisture, rain, and snow (known as rain fade). The system requires a carefully aimed directional antenna.

Satellites in geostationary Earth orbit (GEO) operate in a fixed position 35,786 km (22,236 miles) above the Earth's equator. At the speed of light (about 300,000 km/s or 186,000 miles per second), it takes a quarter of a second for a radio signal to travel from the Earth to the satellite and back. When other switching and routing delays are added and the delays are doubled to allow for a full round-trip transmission, the total delay can be 0.75 to 1.25 seconds. This latency is large when compared to other forms of Internet access with typical latencies that range from 0.015 to 0.2 seconds. Long latencies negatively affect some applications that require real-time response, particularly online games, voice over IP, and remote control devices. TCP tuning and TCP acceleration techniques can mitigate some of these problems. GEO satellites do not cover the Earth's polar regions. HughesNet, Exede, AT&T and Dish Network have GEO systems.

Satellites in low Earth orbit (LEO, below 2000 km or 1243 miles) and medium Earth orbit (MEO, between 2000 and 35,786 km or 1,243 and 22,236 miles) are less common, operate at lower altitudes, and are not fixed in their position above the Earth. Lower altitudes allow lower latencies and make real-time interactive Internet applications more feasible. LEO systems include Globalstar and Iridium. The O3b Satellite Constellation is a proposed MEO system with a latency of 125 ms. COMMStellation™ is a LEO system, scheduled for launch in 2015, that is expected to have a latency of just 7 ms.

Mobile Broadband

Service mark for GSMA

Mobile broadband is the marketing term for wireless Internet access delivered through mobile phone towers to computers, mobile phones (called "cell phones" in North America and South Africa, and "hand phones" in Asia), and other digital devices using portable modems. Some mobile services allow more than one device to be connected to the Internet using a single cellular connection using a process called tethering. The modem may be built into laptop computers, tablets, mobile phones, and other devices, added to some devices using PC cards, USB modems, and USB sticks or dongles, or separate wireless modems can be used.

New mobile phone technology and infrastructure is introduced periodically and generally involves a change in the fundamental nature of the service, non-backwards-compatible transmission technology, higher peak data rates, new frequency bands, wider channel frequency bandwidth in Hertz becomes available. These transitions are referred to as generations. The first mobile data services became available during the second generation (2G).

The download (to the user) and upload (to the Internet) data rates given above are peak or maximum rates and end users will typically experience lower data rates.

WiMAX was originally developed to deliver fixed wireless service with wireless mobility added in 2005. CDPD, CDMA2000 EV-DO, and MBWA are no longer being actively developed.

In 2011, 90% of the world's population lived in areas with 2G coverage, while 45% lived in areas with 2G and 3G coverage.

WiMAX

Worldwide Interoperability for Microwave Access (WiMAX) is a set of interoperable implementations of the IEEE 802.16 family of wireless-network standards certified by the WiMAX Forum. WiMAX enables "the delivery of last mile wireless broadband access as an alternative to cable and DSL". The original IEEE 802.16 standard, now called "Fixed WiMAX", was published in 2001 and provided 30 to 40 megabit-per-second data rates. Mobility support was added in 2005. A 2011 update provides data rates up to 1 Gbit/s for fixed stations. WiMax offers a metropolitan area network with a signal radius of about 50 km (30 miles), far surpassing the 30-metre (100-foot) wireless range of a conventional Wi-Fi local area network (LAN). WiMAX signals also penetrate building walls much more effectively than Wi-Fi.

Wireless ISP

Wi-Fi logo

Wireless Internet service providers (WISPs) operate independently of mobile phone operators. WISPs typically employ low-cost IEEE 802.11 Wi-Fi radio systems to link up remote locations over great distances (Long-range Wi-Fi), but may use other high-er-power radio communications systems as well.

Traditional 802.11b is an unlicensed omnidirectional service designed to span between 100 and 150 m (300 to 500 ft). By focusing the radio signal using a directional antenna 802.11b can operate reliably over a distance of many km(miles), although the technology's line-of-sight requirements hamper connectivity in areas with hilly or heavily foli-ated terrain. In addition, compared to hard-wired connectivity, there are security risks (unless robust security protocols are enabled); data rates are significantly slower (2 to 50 times slower); and the network can be less stable, due to interference from other wireless devices and networks, weather and line-of-sight problems.

Deploying multiple adjacent Wi-Fi access points is sometimes used to create city-wide wireless networks. Some are by commercial WISPs but grassroots efforts have also led to wireless community networks. Rural wireless-ISP installations are typically not commercial in nature and are instead a patchwork of systems built up by hobbyists mounting antennas on radio masts and towers, agricultural storage silos, very tall trees, or whatever other tall objects are available. There are a number of companies that pro-vide this service.

Proprietary technologies like Motorola Canopy & Expedience can be used by a WISP to offer wireless access to rural and other markets that are hard to reach using Wi-Fi or WiMAX.

Local Multipoint Distribution Service

Local Multipoint Distribution Service (LMDS) is a broadband wireless access technol-ogy that uses microwave signals operating between 26 GHz and 29 GHz. Originally designed for digital television transmission (DTV), it is conceived as a fixed wireless, point-to-multipoint technology for utilization in the last mile. Data rates range from 64 kbit/s to 155 Mbit/s. Distance is typically limited to about 1.5 miles (2.4 km), but links of up to 5 miles (8 km) from the base station are possible in some circumstances.

LMDS has been surpassed in both technological and commercial potential by the LTE and WiMAX standards.

Pricing and Spending

Internet access is limited by the relation between pricing and available resources to spend. Regarding the latter, it is estimated that 40% of the world's population has less than US$20 per year available to spend on information and communications technology (ICT). In Mexico, the poorest 30% of the society counts with an estimat-ed US$35 per year (US$3 per month) and in Brazil, the poorest 22% of the popula-tion counts with merely US$9 per year to spend on ICT (US$0.75 per month). From Latin America it is known that the borderline between ICT as a necessity good and

ICT as a luxury good is roughly around the "magical number" of US$10 per person per month, or US$120 per year. This is the amount of ICT spending people esteem to be a basic necessity. Current Internet access prices exceed the available resources by large in many countries.

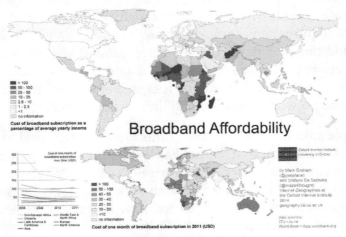

Broadband affordability in 2011

This map presents an overview of broadband affordability, as the relationship between average yearly income per capita and the cost of a broadband subscription (data referring to 2011). Source: Information Geographies at the Oxford Internet Institute.

Dial-up users pay the costs for making local or long distance phone calls, usually pay a monthly subscription fee, and may be subject to additional per minute or traffic based charges, and connect time limits by their ISP. Though less common today than in the past, some dial-up access is offered for "free" in return for watching banner ads as part of the dial-up service. NetZero, BlueLight, Juno, Freenet (NZ), and Free-nets are examples of services providing free access. Some Wireless community networks continue the tradition of providing free Internet access.

Fixed broadband Internet access is often sold under an "unlimited" or flat rate pricing model, with price determined by the maximum data rate chosen by the customer, rather than a per minute or traffic based charge. Per minute and traffic based charges and traffic caps are common for mobile broadband Internet access.

Internet services like Facebook, Wikipedia and Google have built special programs to partner with mobile network operators (MNO) to introduce *zero-rating* the cost for their data volumes as a means to provide their service more broadly into developing markets.

With increased consumer demand for streaming content such as video on demand and peer-to-peer file sharing, demand for bandwidth has increased rapidly and for some ISPs the flat rate pricing model may become unsustainable. However, with fixed costs estimated to represent 80–90% of the cost of providing broadband service, the marginal cost to carry additional traffic is low. Most ISPs do not disclose their costs, but the cost to transmit a gigabyte of data in 2011 was estimated to be about $0.03.

Some ISPs estimate that a small number of their users consume a disproportionate portion of the total bandwidth. In response some ISPs are considering, are experimenting with, or have implemented combinations of traffic based pricing, time of day or "peak" and "off peak" pricing, and bandwidth or traffic caps. Others claim that because the marginal cost of extra bandwidth is very small with 80 to 90 percent of the costs fixed regardless of usage level, that such steps are unnecessary or motivated by concerns other than the cost of delivering bandwidth to the end user.

In Canada, Rogers Hi-Speed Internet and Bell Canada have imposed bandwidth caps. In 2008 Time Warner began experimenting with usage-based pricing in Beaumont, Texas. In 2009 an effort by Time Warner to expand usage-based pricing into the Rochester, New York area met with public resistance, however, and was abandoned. On August 1, 2012 in Nashville, Tennessee and on October 1, 2012 in Tucson, Arizona Comcast began tests that impose data caps on area residents. In Nashville exceeding the 300 Gbyte cap mandates a temporary purchase of 50 Gbytes of additional data.

Digital Divide

Internet users in 2012 as a percentage of a country's population

Fixed broadband Internet subscriptions in 2012 as a percentage of a country's population

Mobile broadband Internet subscriptions in 2012 as a percentage of a country's population

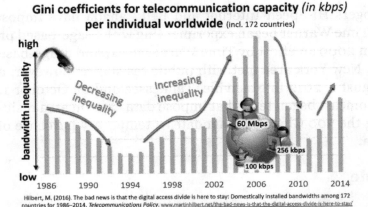

The digital divide measured in terms of bandwidth is not closing, but fluctuating up and down. Gini coefficients for telecommunication capacity (in kbit/s) among individuals worldwide

Despite its tremendous growth, Internet access is not distributed equally within or between countries. The digital divide refers to "the gap between people with effective access to information and communications technology (ICT), and those with very limited or no access". The gap between people with Internet access and those without is one of many aspects of the digital divide. Whether someone has access to the Internet can depend greatly on financial status, geographical location as well as government policies. "Low-income, rural, and minority populations have received special scrutiny as the technological "have-nots."

Government policies play a tremendous role in bringing Internet access to or limiting access for underserved groups, regions, and countries. For example, in Pakistan, which is pursuing an aggressive IT policy aimed at boosting its drive for economic modernization, the number of Internet users grew from 133,900 (0.1% of the population) in 2000 to 31 million (17.6% of the population) in 2011. In countries such as North Korea and Cuba there is relatively little access to the Internet due to the governments' fear of political instability that might accompany the benefits of access to the global Internet. The U.S. trade embargo is another barrier limiting Internet access in Cuba.

Access to computers is a dominant factor in determining the level of Internet access. In 2011, in developing countries, 25% of households had a computer and 20% had Internet access, while in developed countries the figures were 74% of households had a computer and 71% had Internet access. When buying computers was legalized in Cuba in 2007, the private ownership of computers soared (there were 630,000 computers available on the island in 2008, a 23% increase over 2007).

Internet access has changed the way in which many people think and has become an integral part of peoples economic, political, and social lives. The United Nations has recognized that providing Internet access to more people in the world will allow them to take advantage of the "political, social, economic, educational, and career opportunities" available over the Internet. Several of the 67 principles adopted at the World Summit on the Information Society convened by the United Nations in Geneva in 2003, directly address the digital divide. To promote economic development and a reduction of the digital divide, national broadband plans have been and are being developed to increase the availability of affordable high-speed Internet access throughout the world.

Growth in Number of Users

Access to the Internet grew from an estimated 10 million people in 1993, to almost 40 million in 1995, to 670 million in 2002, and to 2.7 billion in 2013. With market saturation, growth in the number of Internet users is slowing in industrialized countries, but continues in Asia, Africa, Latin America, the Caribbean, and the Middle East.

There were roughly 0.6 billion fixed broadband subscribers and almost 1.2 billion mobile broadband subscribers in 2011. In developed countries people frequently use both fixed and mobile broadband networks. In developing countries mobile broadband is often the only access method available.

Bandwidth Divide

Traditionally the divide has been measured in terms of the existing numbers of subscriptions and digital devices ("have and have-not of subscriptions"). Recent studies have measured the digital divide not in terms of technological devices, but in terms of the existing bandwidth per individual (in kbit/s per capita). As shown in the Figure on the side, the digital divide in kbit/s is not monotonically decreasing, but re-opens up with each new innovation. For example, "the massive diffusion of narrow-band Internet and mobile phones during the late 1990s" increased digital inequality, as well as "the initial introduction of broadband DSL and cable modems during 2003–2004 increased levels of inequality". This is because a new kind of connectivity is never introduced instantaneously and uniformly to society as a whole at once, but diffuses slowly through social networks. As shown by the Figure, during the mid-2000s, communication capacity was more unequally distributed than during the late 1980s, when only fixed-line phones existed. The most recent increase in digital equality stems from the massive diffusion of the latest digital innovations (i.e. fixed and mobile broadband infrastructures, e.g. 3G and fiber optics FTTH). As shown in the Figure, Internet access in terms of bandwidth is more unequally distributed in 2014 as it was in the mid-1990s.

In the United States

In the United States, billions of dollars has been invested in efforts to narrow the digital divide and bring Internet access to more people in low-income and rural areas of the United States. Internet availability varies widely state by state in the U.S. In 2011 for example, 87.1% of all New Hampshire residents lived in a household where Internet was available, ranking first in the nation. Meanwhile, 61.4% of all Mississippi residents lived in a household where Internet was available, ranking last in the nation. The Obama administration has continued this commitment to narrowing the digital divide through the use of stimulus funding. The National Center for Education Statistics reported that 98% of all U.S. classroom computers had Internet access in 2008 with roughly one computer with Internet access available for every three students. The percentage and ratio of students to computers was the same for rural schools (98% and 1 computer for every 2.9 students).

Rural Access

One of the great challenges for Internet access in general and for broadband access in particular is to provide service to potential customers in areas of low population density, such as to farmers, ranchers, and small towns. In cities where the population density is high, it is easier for a service provider to recover equipment costs, but each rural customer may require expensive equipment to get connected. While 66% of Americans had an Internet connection in 2010, that figure was only 50% in rural areas, according to the Pew Internet & American Life Project. Virgin Media advertised over 100 towns across the United Kingdom "from Cwmbran to Clydebank" that have access to their 100 Mbit/s service.

Wireless Internet Service Provider (WISPs) are rapidly becoming a popular broadband option for rural areas. The technology's line-of-sight requirements may hamper connectivity in some areas with hilly and heavily foliated terrain. However, the Tegola project, a successful pilot in remote Scotland, demonstrates that wireless can be a viable option.

The Broadband for Rural Nova Scotia initiative is the first program in North America to guarantee access to "100% of civic addresses" in a region. It is based on Motorola Canopy technology. As of November 2011, under 1000 households have reported access problems. Deployment of a new cell network by one Canopy provider (Eastlink) was expected to provide the alternative of 3G/4G service, possibly at a special unmetered rate, for areas harder to serve by Canopy.

A rural broadband initiative in New Zealand is a joint project between Vodafone and Chorus, with Chorus providing the fibre infrastructure and Vodafone providing wireless broadband, supported by the fibre backhaul.

Access as a Civil or Human Right

The actions, statements, opinions, and recommendations outlined below have led to the suggestion that Internet access itself is or should become a civil or perhaps a human right.

Several countries have adopted laws requiring the state to work to ensure that Internet access is broadly available and/or preventing the state from unreasonably restricting an individual's access to information and the Internet:

- Costa Rica: A 30 July 2010 ruling by the Supreme Court of Costa Rica stated: "Without fear of equivocation, it can be said that these technologies [information technology and communication] have impacted the way humans communicate, facilitating the connection between people and institutions worldwide and eliminating barriers of space and time. At this time, access to these technologies becomes a basic tool to facilitate the exercise of fundamental rights and democratic participation (e-democracy) and citizen control, education, freedom of thought and expression, access to information and public services online, the right to communicate with government electronically and administrative transparency, among others. This includes the fundamental right of access to these technologies, in particular, the right of access to the Internet or World Wide Web."

- Estonia: In 2000, the parliament launched a massive program to expand access to the countryside. The Internet, the government argues, is essential for life in the 21st century.

- Finland: By July 2010, every person in Finland was to have access to a one-megabit per second broadband connection, according to the Ministry of Transport and Communications. And by 2015, access to a 100 Mbit/s connection.

- France: In June 2009, the Constitutional Council, France's highest court, declared access to the Internet to be a basic human right in a strongly-worded decision that struck down portions of the HADOPI law, a law that would have tracked abusers and without judicial review automatically cut off network access to those who continued to download illicit material after two warnings

- Greece: Article 5A of the Constitution of Greece states that all persons has a right to participate in the Information Society and that the state has an obligation to facilitate the production, exchange, diffusion, and access to electronically transmitted information.

- Spain: Starting in 2011, Telefónica, the former state monopoly that holds the country's "universal service" contract, has to guarantee to offer "reasonably" priced broadband of at least one megabyte per second throughout Spain.

In December 2003, the World Summit on the Information Society (WSIS) was convened under the auspice of the United Nations. After lengthy negotiations between governments, businesses and civil society representatives the WSIS Declaration of Principles was adopted reaffirming the importance of the Information Society to maintaining and strengthening human rights:

1. We, the representatives of the peoples of the world, assembled in Geneva from 10–12 December 2003 for the first phase of the World Summit on the Information Society, declare our common desire and commitment to build a people-centred, inclusive and development-oriented Information Society, where everyone can create, access, utilize and share information and knowledge, enabling individuals, communities and peoples to achieve their full potential in promoting their sustainable development and improving their quality of life, premised on the purposes and principles of the Charter of the United Nations and respecting fully and upholding the Universal Declaration of Human Rights.

3. We reaffirm the universality, indivisibility, interdependence and interrelation of all human rights and fundamental freedoms, including the right to development, as enshrined in the Vienna Declaration. We also reaffirm that democracy, sustainable development, and respect for human rights and fundamental freedoms as well as good governance at all levels are interdependent and mutually reinforcing. We further resolve to strengthen the rule of law in international as in national affairs.

The WSIS Declaration of Principles makes specific reference to the importance of the right to freedom of expression in the "Information Society" in stating:

4. We reaffirm, as an essential foundation of the Information Society, and as outlined in Article 19 of the Universal Declaration of Human Rights, that everyone has the right to freedom of opinion and expression; that this right includes freedom to hold opinions without interference and to seek, receive and impart information and ideas through any media and regardless of frontiers. Communication is a fundamental social process, a basic human need and the foundation of all social organisation. It is central to the Information Society. Everyone, everywhere should have the opportunity to participate and no one should be excluded from the benefits of the Information Society offers."

A poll of 27,973 adults in 26 countries, including 14,306 Internet users, conducted for the BBC World Service between 30 November 2009 and 7 February 2010 found that almost four in five Internet users and non-users around the world felt that access to the Internet was a fundamental right. 50% strongly agreed, 29% somewhat agreed, 9% somewhat disagreed, 6% strongly disagreed, and 6% gave no opinion.

The 88 recommendations made by the Special Rapporteur on the promotion and protection of the right to freedom of opinion and expression in a May 2011 report to the Human Rights Council of the United Nations General Assembly include several that bear on the question of the right to Internet access:

> 67. Unlike any other medium, the Internet enables individuals to seek, receive and impart information and ideas of all kinds instantaneously and inexpensively across national borders. By vastly expanding the capacity of individuals to enjoy their right to freedom of opinion and expression, which is an "enabler" of other human rights, the Internet boosts economic, social and political development, and contributes to the progress of humankind as a whole. In this regard, the Special Rapporteur encourages other Special Procedures mandate holders to engage on the issue of the Internet with respect to their particular mandates.

> 78. While blocking and filtering measures deny users access to specific content on the Internet, States have also taken measures to cut off access to the Internet entirely. The Special Rapporteur considers cutting off users from Internet access, regardless of the justification provided, including on the grounds of violating intellectual property rights law, to be disproportionate and thus a violation of article 19, paragraph 3, of the International Covenant on Civil and Political Rights.

> 79. The Special Rapporteur calls upon all States to ensure that Internet access is maintained at all times, including during times of political unrest.

> 85. Given that the Internet has become an indispensable tool for realizing a range of human rights, combating inequality, and accelerating development and human progress, ensuring universal access to the Internet should be a priority for all States. Each State should thus develop a concrete and effective policy, in consultation with individuals from all sections of society, including the private sector and relevant Government ministries, to make the Internet widely available, accessible and affordable to all segments of population.

Network Neutrality

Network neutrality (also net neutrality, Internet neutrality, or net equality) is the principle that Internet service providers and governments should treat all data on the Internet equally, not discriminating or charging differentially by user, content, site, platform, application, type of attached equipment, or mode of communication. Advocates of net neutrality have raised concerns about the ability of broadband providers to use their last mile infrastructure to block Internet applications and content (e.g. websites, services, and protocols), and even to block out competitors. Opponents claim net neutrality regulations would deter investment into improving broadband infrastructure and try to fix something that isn't broken.

Natural Disasters and Access

Natural disasters disrupt internet access in profound ways. This is important—not only for telecommunication companies who own the networks and the businesses who use them, but for emergency crew and displaced citizens as well. The situation is worsened when hospitals or other buildings necessary to disaster response lose their connection. Knowledge gained from studying past internet disruptions by natural disasters could be put to use in planning or recovery. Additionally, because of both natural and man-made disasters, studies in network resiliency are now being conducted to prevent large-scale outages.

One way natural disasters impact internet connection is by damaging end sub-networks (subnets), making them unreachable. A study on local networks after Hurricane Ka-trina found that 26% of subnets within the storm coverage were unreachable. At Hurricane Katrina's peak intensity, almost 35% of networks in Mississippi were without power, while around 14% of Louisiana's networks were disrupted. Of those unreachable subnets, 73% were disrupted for four weeks or longer and 57% were at "network edges where important emergency organizations such as hospitals and government agencies are mostly located". Extensive infrastructure damage and inaccessible areas were two explanations for the long delay in returning service. The company Cisco has revealed a Network Emergency Response Vehicle (NERV), a truck that makes portable communications possible for emergency responders despite traditional networks being disrupted.

A second way natural disasters destroy internet connectivity is by severing submarine cables—fiber-optic cables placed on the ocean floor that provide international internet connection. The 2006 undersea earthquake near Taiwan (Richter scale 7.2) cut six out of seven international cables connected to that country and caused a tsunami that wiped out one of its cable and landing stations. The impact slowed or disabled internet connection for five days within the Asia-Pacific region as well as between the region and the United States and Europe.

With the rise in popularity of cloud computing, concern has grown over access to cloud-hosted data in the event of a natural disaster. Amazon Web Services (AWS) has been in the news for major network outages in April 2011 and June 2012. AWS, like other major cloud hosting companies, prepares for typical outages and large-scale natural disasters with backup power as well as backup data centers in other locations. AWS divides the globe into five regions and then splits each region into availability zones. A data center in one availability zone should be backed up by a data center in a different availability zone. Theoretically, a natural disaster would not affect more than one availability zone. This theory plays out as long as human error is not added to the mix. The June 2012 major storm only disabled the primary data center, but human error disabled the secondary and tertiary backups, affecting companies such as Netflix, Pinterest, Reddit, and Instagram.

Triple Play (Telecommunications)

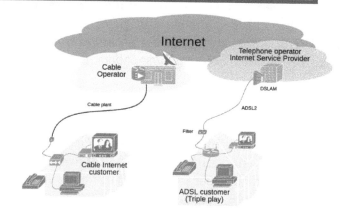

In telecommunications, triple play service is a marketing term for the provisioning, over a single broadband connection, of: two bandwidth-intensive services, broadband Internet access and television, and the latency-sensitive telephone. Triple play focuses on a supplier convergence rather than solving technical issues or a common standard. However, standards like G.hn might deliver all these services on a common technology.

Quadruple Play

A so-called quadruple play (or quad play) service integrates mobility as well, often by supporting dual mode mobile plus hotspot-based phones that shift from GSM to WiFi when they come in range of a home wired for triple-play service. Typical Generic Access Network services of this kind, such as Rogers Home Calling Zone (Rogers is an incumbent in the Canadian market), allow the caller to enter and leave the range of their home Wi-Fi network, and only pay GSM rates for the time they spend outside the range. Calls at home are routed over the IP network and paid at a flat rate per month. No interruption or authorization for the shift is required—soft handoff takes place automatically as many times as the caller enters or leaves the range.

CATV

By about 2000, cable TV companies were in a technical position to offer *triple play* over one physical medium to a large number of their customers, as their networks already have sufficient bandwidth to carry hundreds of video channels. Cable's main competition for television in North America came from satellites, which cannot compete for voice and interactive broadband due to the latency imposed by physical laws on a geo-synchronous satellite—sometimes up to one full second of delay between speaking and being heard. Cable's main competition for voice and Internet access came from telcos, which were not yet able to compete for television in most markets because DSL over most local loops could not provide enough bandwidth.

As an interim marketing move while they installed fiber closer to the customer, telcos such as AT&T did co-promotion deals with satellite TV providers to sell television, telephone, and Internet access services bundled for billing purposes although the services provided through a satellite link and the services provided through a phone line are not technically related. Telcos that own wireless phone networks also included those as part of such billing-only bundles because most cable companies do not own wireless networks

Deployments

The first triple-play deployment was by Italian operator Fastweb in 2001, using fibre to the home service and one of the first triple-play home gateway devices with embedded fibre termination. This enabled the operator to deliver voice, video, and data services to subscribers' homes via its 10 MB SDSL network. This approach, known as Point-to-Point Protocol over Ethernet. This FTTH architecture brought the operator the best ARPU in the industry for a number of consecutive years.

Triple-play services in the United States are offered by cable television operators as well as by telecommunication operators, who directly compete with one another. Providers expect that an integrated solution will increase opportunity costs for customers who may want to choose between service providers, permit more cross-selling, and hold off the power companies deploying G.hn and IEEE P1901 technology with its radically superior service and deployment characteristics for at least another decade or so.

Outside the United States, notably in Ecuador, Pakistan, India, Japan, and China, power companies have generally been more successful in leapfrogging legacy technologies. In Switzerland and Sweden, dark fiber is available reliably to homes at tariffed rates (in Switzerland four dark fibres are deployed to each home) supporting speeds in excess of 40 Gbit/s—only to the local caches, however, as backhaul cannot typically support more than 10 Mbit/s connections to global services.

Since 2007, access providers in Italy have been participating in an initiative called Fiber for Italy, which aims to build an infrastructure that can deliver 100 Mbit/s symmetrical bandwidth to consumers, in order to enable the delivery of triple- and quad-play services.

Other triple-play deployments include Deutsche Telekom, Telecom Italia, Swisscom, Telekom Austria, and Telus.

Regulation

There are multiple and intense regulatory battles over triple-play services, as incumbent telcos and incumbent cable operators attempt to keep out new competitors; since both industries historically have been regulated monopolies, regulatory capture has long been as much a core competency for them as have been prices and terms of service. Cable providers want to compete with telcos for local voice service, but want to discourage telcos from competing with them for television service. Incumbent telcos want to deliver television service but want to block competition for voice service from cable operators. Both industries cloak their demands for favorable regulatory treatment in claims that their positions favor the public interests.

In March 2007 cable operators scored a major victory when the Federal Communications Commission (FCC) overruled two state public service commissions by ruling that incumbent local exchange carriers must connect to VoIP services . Regulators in South Carolina and Nebraska had been allowing local telcos to block Time Warner Cable from offering local phone service in their states. In the other direction, also in March 2007, the FCC limited the powers of municipalities and states over telcos that want to compete with cable TV companies. Consumer groups expressed displeasure with this FCC ruling because they fear telcos will only offer service to the richest neighborhoods, which is a major point of contention between telcos wanting to offer television service and local governments is that local governments typically impose "build-out" and community access requirements so a cable provider is forced to wire the entire town within a specified period of time. All three Republican members of the FCC voted for this decision, while both Democratic members voted against it and one predicted either U.S. Congress or the courts would overturn it. In October 2007, The *Hartford Courant* reported that Connecticut regulators have ordered AT&T to stop signing up new customers for its IPTV service until they got a cable license; AT&T said they would fight this decision in court .

Telco

For telephone local exchange carriers (LEC), *triple play* is delivered using a combination of optical fiber and digital subscriber line (DSL) technologies (called fiber in the loop) to its residential base. This configuration uses fiber communications to reach distant locations and uses DSL over an existing POTS twisted pair cable as last mile access to the subscriber's home. Cable television operators use a similar architecture called hybrid fibre coaxial (HFC) to provide subscriber homes with broadband, but use the available coaxial cable rather than a twisted pair for the last mile transmission standard. Subscriber homes can be in a residential environment, multi-dwelling units, or even in business offices.

Using DSL over twisted pair, television content is delivered using IPTV where the content is streamed to the subscriber in an MPEG-2 transport format. On an HFC network, television may be a mixture of analog and digital television signals. A set-top box (STB) is used at the subscriber's home to allow the susbcriber to control viewing and order new video services such as "movies on demand". Access to the Internet is provided through ATM or DOCSIS, typically provided as an Ethernet port to the subscriber. Voice service can be provided using a traditional plain old telephone service (POTS) interface as part of the legacy telephone network or can be delivered using voice over IP (VoIP). In an HFC network, voice is delivered using VoIP.

Some service providers are also rolling out Ethernet to the home networks and fiber to the home, which support triple-play services and bypass the disadvantages of adapting broadband transmission to a legacy network. This is particularly common in greenfield developments where the capital expenditure is reduced by deploying one network to deliver all services.

For existing multiple-dwelling-unit (MDU) buildings, where running fiber to each unit may not be feasible, telcos often use VDSL to connect individual units over existing copper through a central optical network terminal located in the existing telco closet. Over such a short distance, DSL can deliver much higher bitrates than is possible running DSL over the local loop from the nearest central office (as is common with basic DSL).

Wireless

Triple play has led to the term "quadruple play", where wireless communications is introduced as another medium to deliver video, Internet access, and voice telephone service. Advances in both CDMA and GSM standards, utilizing 3G, 4G, or UMTS allows the service operators to enter into quadruple play and gain competitive advantage against other providers. The grouping together of services (as triple or quadruple play) is called multi-play.

Other advanced technologies such as WiMax or 802.16 has allowed new market entrants to achieve triple play. Many speculate that this means serious, new competition for established providers of bundled telecommunications services.

Power Integration

These services can be delivered with a BPL network using technologies such as IEEE P1901/G.hn. Since the devices all rely on AC power (or DC power via 802.3af or 802.3at which rely on AC power at the PoE hub), connecting them with only one cable each for both power and gigabit data cuts wiring costs, and most rooms are already wired for power.

Business

The challenges in offering *triple play* are mostly associated with determining the right business model, backend processes, customer care support, and economic environment, rather than technology. For example, using the right billing platform to address a variety of subscriber demographics or having the appropriate subscriber density to financially justify introduction of the service are a few factors that affect decisions to offer triple play.

In addition to the challenges mentioned above, there are a number of technical challenges with regards to the rollout of triple-play services. Voice, video, and high-speed data all have different characteristics and place different burdens on the network that provides access to these services. Voice services are greatly affected by jitter, whereas packet loss has a greater effect on video and data services. In order to use a shared network resource such as cable or DSL, the service provider may use network equipment that employs quality-of-service mechanisms to adjust to the requirements of the different services.

References

- William Stallings (1999). ISDN and Broadband ISDN with Frame Relay and ATM (4th ed.). Prentice Hall. p. 542. ISBN 0139737448.

- Telecommunications and Data Communications Handbook, Ray Horak, 2nd edition, Wiley-Interscience, 2008, 791 p., ISBN 0-470-39607-5

- Dean, Tamara (2009). Network+ Guide to Networks (5th ed.). Course Technology, Cengage Learning. ISBN 1-4239-0245-9. pp 312–315.

- Berger, Lars T.; Schwager, Andreas; Pagani, Pascal; Schneider, Daniel M. (February 2014). MIMO Power Line Communications: Narrow and Broadband Standards, EMC, and Advanced Processing. Devices, Circuits, and Systems. CRC Press. ISBN 9781466557529.

- Joseph N. Pelton (2006). The Basics of Satellite Communication. Professional Education International, Inc. ISBN 978-1-931695-48-0.

- Deborah Hurley, James H. Keller (1999). The First 100 Feet: Options for Internet and Broadband Access. Harvard college. ISBN 0-262-58160-4.

- Mustafa Ergen (2009). Mobile Broadband: including WiMAX and LTE. Springer Science+Business Media. ISBN 978-0-387-68189-4.

- Joshua Bardwell; Devin Akin (2005). Certified Wireless Network Administrator Official Study Guide (Third ed.). McGraw-Hill. p. 418. ISBN 978-0-07-225538-6.

- Wattal, S.; Yili Hong; Mandviwalla, M.; Jain, A., "Technology Diffusion in the Society: Analyzing Digital Divide in the Context of Social Class", Proceedings of the 44th Hawaii International Conference on System Sciences (HICSS), pp.1–10, 4–7 January 2011, ISBN 978-0-7695-4282-9

- "Changes in Cuba: From Fidel to Raul Castro", Perceptions of Cuba: Canadian and American policies in comparative perspective, Lana Wylie, University of Toronto Press Incorporated, 2010, p. 114, ISBN 978-1-4426-4061-0

- Ji, Dagyum (2016-08-18). "Netflix style video-on-demand comes to North Korea, state TV shows". NK News. Retrieved 2016-08-25.

- Ben Munson (June 29, 2016). "Akamai: Global average internet speeds have doubled since last Olympics". FierceOnlineVideo. Retrieved June 30, 2016.

5

Applications of WiMax Technology

WiMAX has found use in wireless backhaul technology, hotspot and wireless LAN networks. This chapter studies these applications of WiMAX and makes the reader familiar with how WiMAX has rendered bigger prospects to these technologies and networks. This chapter discusses the methods of WiMAX in a critical manner providing key analysis to the subject matter.

Backhaul (Telecommunications)

In a hierarchical telecommunications network the backhaul portion of the network comprises the intermediate links between the core network, or backbone network and the small subnetworks at the "edge" of the entire hierarchical network.

In contracts pertaining to such networks, backhaul is the obligation to carry packets to and from that global network. A non-technical business definition of *backhaul* is the commercial wholesale bandwidth provider who offers quality of service (QOS) guarantees to the retailer. It appears most often in telecommunications trade literature in this sense, whereby the backhaul connection is defined not technically but by who operates and manages it, and who takes legal responsibility for the connection or uptime to the Internet or 3G/4G network.

In both the technical and commercial definitions, backhaul generally refers to the side of the network that communicates with the global Internet, paid for at wholesale commercial access rates to or at an ethernet exchange or other core network access location. Sometimes middle mile networks exist between the customer's own LAN and those exchanges. This can be a local WAN or WLAN connection, for instance Network New Hampshire Now and Maine Fiber Company run tariffed public dark fiber networks as a backhaul alternative to encourage local and national carriers to reach areas with broadband and cell phone that they otherwise would not be serving. These serve retail networks which in turn connect buildings and bill customers directly.

Cell phones communicating with a single cell tower constitute a local subnetwork; the connection between the cell tower and the rest of the world begins with a backhaul link to the core of the Internet service provider's network (via a point of presence). The term backhaul may be used to describe the entire wired part of a network, although some networks have wireless instead of wired backhaul, in whole or in part, for example using microwave bands and mesh network and edge network topologies that may use a high-capacity wireless channel to get packets to the microwave or fiber links.

A telephone company is very often the ISP providing backhaul, although for academic R&E networks or large commercial networks or municipal networks, it is increasingly common to connect to a public broadband backhaul. See national broadband plans from around the world, many of which were motivated by the perceived need to break the monopoly of incumbent commercial providers. The US plan for instance specifies that all community anchor institutions should be connected by gigabit fiber optics before the end of 2020.

Definition

Visualizing the entire hierarchical network as a human skeleton, the core network would represent the spine, the backhaul links would be the limbs, the edge networks would be the hands and feet, and the individual links within those edge networks would be the fingers and toes.

Other examples include:

- Connecting wireless base stations to the corresponding base station controllers.

- Connecting DSLAMs to the nearest ATM or Ethernet aggregation node.

- Connecting a large company's site to a metro Ethernet network.

- Connecting a submarine communications cable system landing point (which is usually in a remote location) with the main terrestrial telecommunications network of the country that the cable serves.

Available Backhaul Technologies

CableFree Microwave Backhaul links deployed for mobile operators in the Middle East. These microwave links typically carry a mix of Ethernet /IP, TDM (Nx E1) and SDH traffic to connect the Cellular Base Stations (BTS) to the central sites of the cellular operator. Such microwave links used to carry 2xE1 (4Mbit/s) now carry 400Mbit/s or more, using modern 1024QAM or higher modulation schemes.

The choice of backhaul technology must take account of such parameters as capacity, cost, reach, and the need for such resources as frequency spectrum, optical fiber, wiring, or rights of way.

Generally, backhaul solutions can largely be categorised into wired (leased lines or copper/fibre) or wireless (point-to-point, point-to-multipoint over high-capacity radio links). Wired is usually a very expensive solution and often impossible to deploy in remote areas, hence making wireless a more suitable and/or a viable option. Multi-hop wireless architecture can overcome the hurdles of wired solutions to create efficient large coverage areas and with growing demand in emerging markets where often cost is a major factor in deciding technologies, a wireless backhaul solution is able to offer 'carrier-grade' services, whereas this is not easily feasible with wired backhaul connectivity.

Backhaul technologies include:

- Free space optics (FSO)

- Point-to-point microwave radio relay transmission (terrestrial or, in some cases, by satellite)

- Point-to-multipoint microwave-access technologies, such as LMDS, Wi-Fi, Wi-MAX, etc., can also function for backhauling purposes

- DSL variants, such as ADSL, VDSL and SHDSL

- PDH and SDH/SONET interfaces, such as (fractional) E1/T1, E3, T3, STM-1/OC-3, etc.

- Ethernet

Backhaul capacity can also be leased from another network operator, in which case that other network operator generally selects the technology being used, though this can be limited to fewer technologies if the requirement is very specific such as short-term links for emergency/disaster relief or for public events, where cost and time would be major factors and would immediately rule out wired solutions, unless pre-existing infrastructure was readily accessible or available.

WiFi Mesh Networks for Wireless Backhaul

As data rates increase, the range of wireless network coverage is reduced, raising investment costs for building infrastructure with access points to cover service areas. Mesh networks are unique enablers that can reduce this cost due to their flexible architecture.

With mesh networking, access points are connected wirelessly and exchange data frames with each other to forward to/from a gateway point.

Since a mesh requires no costly cable constructions for its backhaul network, it reduces total investment cost. Mesh technology's capabilities can boost extending coverage of service areas easily and flexibly.

For further cost reduction, a large-scale high-capacity mesh is desirable. For instance, Kyushu University's Mimo-Mesh Project, based in Fukuoka City, Fukuoka Prefecture, Japan, has developed and put into use new technology for building high capacity mesh infrastructure. A key component is called IPT, intermittent periodic transmit, a proprietary packet-forwarding scheme that is designed to reduce radio interference in the forwarding path of mesh networks. In 2010, hundreds of wireless LAN access points incorporating the technology were installed in the commercial shopping and entertainment complex, Canal City Hakata, resulting in the successful operation of one of the world's largest indoor wireless multi-hop backhauls. That network uses a wireless multi-hop relay of up to 11 access points while delivering high bandwidth to end users. Actual throughput is double that of standard mesh network systems using conventional packet forwarding. Latency, as in all multi-hop relays, suffers, but not to the degree that it compromises voice over IP communications.

Open Solutions: Using Many Connections as a Backhaul

Many common wireless mesh network hotspot solutions are supported in open source router firmware including DD-WRT, OpenWRT and derivatives. Sputnik Agent, Hotspot System, Chillispot and the ad-supported AnchorFree are four examples that work even with lower end routers like the WRT54G. The IEEE 802.21 standard specifies basic capabilities for such systems including 802.11u unknown user authentication and 802.11s ad hoc wireless mesh networking support. Effectively these allow arbitrary wired net connections to be teamed or ganged into what appears to be a single backhaul – a "virtual private cloud". Proprietary networks from Meraki follow similar principles. The use of the term backhaul to describe this type of connectivity may be controversial technically. They invert the business definition, as it is the customer who is providing the connectivity to the open Internet while the vendor is providing authentication and management services.

Transition from Wireless to Wired Backhaul

Wireless backhaul is easy to deploy, and allows moving points of presence, however, these wireless connections are slower, occupy spectrum that could be used by user devices (especially as 5.8 GHz devices proliferate), require more truck rolls (typically three times as many) as wired backhaul, and are limited in bandwidth. They are often viewed as an initial or temporary measure.

Exception: Microwave Superior to Copper in Some High-tower Applications

The exception is very high towers, including cell towers. Where it can be deployed, "microwave is cheaper, more scalable and more reliable than copper, though not as desirable as fiber: "If rural telcos think they can get away with not upgrading from copper, they will just lose [the tower] to microwave." [error: quote is not in source]

Wiring Issues

Strategies for moving from wireless to wired backhaul usually involve building out wired connections only as necessary to improve performance (especially latency). This can often be a simple quantitative analysis: How much spectrum is freed for user purposes, how much better round trip delay (latency) will be achieved, how much more bandwidth can move, and how well does the network respond to stress conditions notably when there are too many requests for connections than a wireless backhaul can accommodate. According to some sources, notably Forbes magazine, for 3G/4G/LTE networks, the

> "scarcity of wide-area spectrum will cause a significant migration towards more local area networks such as femtocells (small, lower-power radio transmission stations) and wifi, and will eventually find a relay in the infinitely expandable, wired backhaul—the link to a provider's core network. Wired fiber infrastructure can still carry vastly more data than any wireless system."

Hardware Issues

Aside from changes to the network (wires and switching and management) a well designed future-proof wireless network may not require much change at the endpoints. All equipment used in a wireless backhaul configuration supports simultaneous dual band communication (one band for user communication, another for the backhaul). Almost all such equipment also supports wired connections, typically using power over Ethernet or (less often in outdoor applications) unpowered Ethernet. So assuming reasonable software flexibility, both bands can be repurposed to user connectivity while the backhaul shifts to a wired connection. Unless nodes are entirely self-powered, they are connected at least indirectly to an AC or DC power source, suggesting that the single-cable solutions (power over Ethernet DC and IEEE P1901 AC) for both power and data should dominate for low-power shorter-range nodes, especially municipal Wi-Fi projects in low usage rural areas.

Software Issues

Software-defined networking (SDN) is also easing the transition between different link layer technologies. Most notably, OpenRadio/OpenFlow software architecture gangs many backhaul sources and makes near-optimal use of all-copper existing-wire infrastructure. As of 2012 these technologies were mostly used in data centres, but wide-area carriers like Verizon were announcing their support for them in customer networks, and many other companies were involved in what was sometimes called the "SDN revolution".

These carrier and data centre initiatives are part of a general trend to redundant back-haul or hybrid networking in which there is more of a lattice than hierarchy of back-haul. For instance, the IEEE P1905 standard permits, and the IEEE 802.21 standard supports in applications, similar connection of multiple connections at the LAN level.

Both standards enable hybrid networks that allow many devices to connect via many protocols rather than being tied to a single backhaul associated with that device or an account solely associated with that device. *cloud computing for commercial and industrial precedents for this consumer-level technology.*

Cell Towers Moving from Microwave to Fiber Optic

Longer range, higher power nodes, however, including cell phone towers, must be direct-ly connected to fiber optic, increasingly using the Carrier Ethernet protocols, replacing older T-1 connections. "Research firm NPD In-Stat projects that by 2014 Ethernet will be the dominant technology for wireless backhaul, with 85% usage in base stations...the momentum is all on the Ethernet side." 50-100 Mbit/s Ethernet circuits are standard.

> "AT&T and Verizon are both very clear that they will only accept fiber to cell towers."

Very Long Range (Including Submarine) Networks

On very large scale long range networks, including transcontinental, submarine tele-communications cables are used. Sometimes these are laid alongside HVDC cables on the same route. Several companies, including Prysmian, run both HVDC power cables and telecommunications cables as far as FTTx. This reflects the fact that telecommuni-cations backhaul and long range high voltage electricity transmission have many tech-nologies in common, and are almost identical in terms of route clearing, liability in outages, and other legal aspects.

Hotspot (Wi-Fi)

A hotspot is a physical location where people may obtain Internet access, typically us-ing Wi-Fi technology, via a wireless local area network (WLAN) using a router connect-ed to an internet service provider.

Public hotspots may be found in a number of businesses for use of customers in many developed urban areas throughout the world, such as coffee shops. Many hotels offer wifi access to guests, either in guest rooms or in the lobby. Hotspots differ from wire-less access points, which are the hardware devices used to provide a wireless network service. Private hotspots allow Internet access to a device (such as a tablet) via another device which may have data access via say a mobile device.

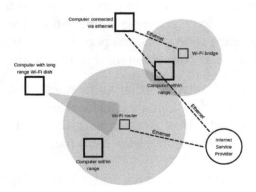

A diagram showing a Wi-Fi network

History

Public access wireless local area networks (LANs) were first proposed by Henrik Sjö-din at the NetWorld+Interop conference in The Moscone Center in San Francisco in August 1993. Sjödin did not use the term hotspot but referred to publicly accessible wireless LANs.

Public park in Brooklyn, New York, has free Wi-Fi from a local corporation

The first commercial venture to attempt to create a public local area access network was a firm founded in Richardson, Texas known as PLANCOM (Public Local Area Network Communications). The founders of the venture, Mark Goode, Greg Jackson, and Brett Stewart dissolved the firm in 1998, while Goode and Jackson created MobileStar Networks. The firm was one of the first to sign such public access locations as Starbucks, American Airlines, and Hilton Hotels. The company was sold to Deutsche Telecom in 2001, who then converted the name of the firm into "T-Mobile Hotspot." It was then that the term "hotspot" entered the popular vernacular as a reference to a location where a publicly accessible wireless LAN is available.

ABI Research reported there was a total of 4.9 million global Wi-Fi hotspots in 2012 and projected that number would surpass 6.3 million by the end of 2013. The latest Wireless Broadband Alliance (WBA) Industry Report outlines a positive scenario for the Wi-Fi market: a steady annual increase from 5.2m public hotspots in 2012 to 10.5m public hotspots in 2018. Collectively, WBA operator members serve more than 1 billion subscribers and operate more than 15 million hotspots globally.

Uses

The public can use a laptop or other suitable portable device to access the wireless connection (usually Wi-Fi) provided. Of the estimated 150 million laptops, 14 million PDAs, and other emerging Wi-Fi devices sold per year for the last few years, most include the Wi-Fi feature.

For venues that have broadband Internet access, offering wireless access is as simple as configuring one access point (AP), in conjunction with a router and connecting the AP to the Internet connection. A single wireless router combining these functions may suffice.

The iPass 2014 interactive map, that shows data provided by the analysts Maravedis Rethink, shows that in December 2014 there are 46,000,000 hotspots worldwide and more than 22,000,000 roamable hotspots. More than 10,900 hotspots are on trains, planes and airports (Wi-Fi in motion) and more than 8,500,000 are "branded" hotspots (retail, cafés, hotels). The region with the largest number of public hotspots is Europe, followed by North America and Asia.

Security

Security is a serious concern in connection with Hotspots. There are three possible attack vectors. First, there is the wireless connection between the client and the access point. This needs to be encrypted, so that the connection cannot be eavesdropped or attacked by a man-in-the-middle-attack. Second, there is the Hotspot itself. The WLAN encryption ends at the interface, then travels its network stack unencrypted and then travels over the wired connection up to the BRAS of the ISP. Third, there is the connection from the Access Point to the BRAS of the ISP.

The safest method when accessing the Internet over a Hotspot, with unknown security measures, is end-to-end encryption. Examples of strong end-to-end encryption are HTTPS and SSH.

Locations

Hotspots are often found at airports, bookstores, coffee shops, department stores, fuel stations, hotels, hospitals, libraries, public pay phones, restaurants, RV parks and campgrounds, supermarkets, train stations, and other public places. Additionally, many schools and universities have wireless networks in their campuses.

Types

Free hotspots operate in two ways:

- Using an open public network is the easiest way to create a free hotspot. All that is needed is a Wi-Fi router. Similarly, when users of private wireless routers turn off their authentication requirements, opening their connection, intentionally or not, they permit piggybacking (sharing) by anyone in range.

- Closed public networks use a HotSpot Management System to control access to hotspots. This software runs on the router itself or an external computer allowing operators to authorize only specific users to access the Internet. Providers of such hotspots often associate the free access with a menu, membership, or purchase limit. Operators may also limit each user's available bandwidth (upload and download speed) to ensure that everyone gets a good quality service. Often this is done through service-level agreements.

Commercial Hotspots

A commercial hotspot may feature:

- A captive portal / login screen / splash page that users are redirected to for authentication and/or payment. The captive portal / splash page sometimes includes the social login buttons.

- A payment option using a credit card, iPass, PayPal, or another payment service (voucher-based Wi-Fi)

- A walled garden feature that allows free access to certain sites

- Service-oriented provisioning to allow for improved revenue

- Data analytics and data capture tools, to analyze and export data from Wi-Fi clients

Many services provide payment services to hotspot providers, for a monthly fee or commission from the end-user income. For example, Amazingports can be used to set up hotspots that intend to offer both fee-based and free internet access, and ZoneCD is a Linux distribution that provides payment services for hotspot providers who wish to deploy their own service.

Major airports and business hotels are more likely to charge for service, though most hotels provide free service to guests; and increasingly, small airports and airline lounges offer free service.. Retail shops, public venues and offices usually provide a free Wi-Fi SSID for their guests and visitors.

Roaming services are expanding among major hotspot service providers. With roaming service the users of a commercial provider can have access to other providers' hotspots, either free of charge or for extra fees, which users will usually be charged on an access-per-minute basis.

Software Hotspots

Many Wi-Fi adapters built into or easily added to consumer computers and mobile devices include the functionality to operate as private or mobile hotspots, sometimes referred to as "mi-fi". The use of a private hotspot to enable other personal devices to access the WAN (usually but not always the Internet) is a form of bridging, and known as tethering. Manufacturers and firmware creators can enable this functionality in Wi-Fi devices on many Wi-Fi devices, depending upon the capabilities of the hardware, and most modern consumer operating systems, including Android, Apple OS X 10.6 and later, Windows mobile, and Linux include features to support this. Additionally wireless chipset manufacturers such as Atheros, Broadcom, Intel and others, may add the capability for certain Wi-Fi NICs, usually used in a client role, to also be used for hotspot purposes. However, some service providers, such as AT&T, Sprint, and T-Mobile charge users for this service or prohibit and disconnect user connections if tethering is detected.

Third-party software vendors offer applications to allow users to operate their own hotspot, whether to access the Internet when on the go, share an existing connection, or extend the range of another hotspot. Third party implementations of software hotspots include:

- AmazingPorts Hotspot software
- Antamedia HotSpot software
- Connectify Hotspot
- Jaze Hotspot Gateway by Jaze Networks
- Hot Spot Network Manager (HSNM)
- Virtual Router
- Tanaza
- Start Hotspot software

Hotspot 2.0

Hotspot 2.0, also known as HS2 and Wi-Fi Certified Passpoint, is an approach to public access Wi-Fi by the Wi-Fi Alliance. The idea is for mobile devices to automatically join a Wi-Fi subscriber service whenever the user enters a Hotspot 2.0 area, in order to provide better bandwidth and services-on-demand to end-users and relieve carrier infrastructure of some traffic.

Hotspot 2.0 is based on the IEEE 802.11u standard, which is a set of protocols published in 2011 to enable cellular-like roaming. If the device supports 802.11u and is subscribed to a Hotspot 2.0 service it will automatically connect and roam.

Supported Devices

- Some Chinese tablet computers

- Some THL smartphones

- Apple mobile devices running iOS 7 and up

- Some Samsung Galaxy smartphones

- Windows 10 devices have full support for network discovery and connection

- Windows 8 and Windows 8.1 lack network discovery, but supports connecting to a network when the credentials are known

Billing

The so-called "User-Fairness-Model " is a dynamic billing model, which allows a volume-based billing, charged only by the amount of payload (data, video, audio). Moreover, the tariff is classified by net traffic and user needs (Pommer, p. 116ff).

If the net traffic increases, then the user has to pay the next higher tariff class. By the way the user is asked for if he still wishes the session also by a higher traffic class.Moreover, in time-critical applications (video, audio) a higher class fare is charged, than for non time-critical applications (such as reading Web pages, e-mail).

The "User-fairness model" can be implemented with the help of EDCF (IEEE 802.11e). A EDCF user priority list shares the traffic in 3 access categories (data, video, audio) and user priorities (UP) (Pommer, p. 117):

- Data [UP 0|2]

- Video [UP 5|4]

- Audio [UP 7|6]

Service-oriented provisioning for viable implementations

Security Concerns

Some hotspots authenticate users; however, this does not prevent users from viewing network traffic using packet sniffers.

Some vendors provide a download option that deploys WPA support. This conflicts with enterprise configurations that have solutions specific to their internal WLAN.

In order to provide robust security to hotspot users, the Wi-Fi Alliance is developing a new hotspot program that aims to encrypt hotspot traffic with WPA2 security. The program was scheduled to launch in the first half of 2012.

Legal Concerns

Depending upon the location, providers of public hotspot access may have legal obligations, related to privacy requirements and liability for use for unlawful purposes. In countries where the internet is regulated or freedom of speech more restricted, there may be requirements such as licensing, logging, or recording of user information. Concerns may also relate to child safety, and social issues such as exposure to objectionable content, protection against cyberbullying and illegal behaviours, and prevention of perpetration of such behaviors by hotspot users themselves.

European Union

- Data Retention Directive Hotspot owners must retain key user statistics for 12 months.

- Directive on Privacy and Electronic Communications

United Kingdom

- Data Protection Act 1998 The hotspot owner must retain individual's information within the confines of the law.

- Digital Economy Act 2010 Deals with, among other things, copyright infringement, and imposes fines of up to £250,000 for contravention.

Wireless LAN

This notebook computer is connected to a wireless access point using a PC card wireless card.

A wireless local area network (WLAN) is a wireless computer network that links two or more devices using a wireless distribution method (often spread-spectrum or OFDM radio) within a limited area such as a home, school, computer laboratory, or office building. This gives users the ability to move around within a local coverage area and yet still be connected to the network. A WLAN can also provide a connection to the wider Internet.

Most modern WLANs are based on IEEE 802.11 standards and are marketed under the Wi-Fi brand name.

Wireless LANs have become popular for use in the home, due to their ease of installation and use. They are also popular in commercial complexes that offer wireless access to their customers (often without charge). New York City, for instance, has begun a pilot program to provide city workers in all five boroughs of the city with wireless Internet access.

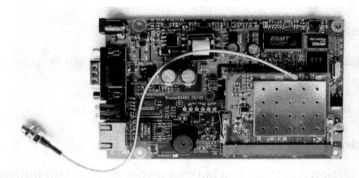

An embedded RouterBoard 112 with U.FL-RSMA pigtail and R52 mini PCI Wi-Fi card widely used by wireless Internet service providers (WISPs)

History

Norman Abramson, a professor at the University of Hawaii, developed the world's first wireless computer communication network, ALOHAnet (operational in 1971), using low-cost ham-like radios. The system included seven computers deployed over four islands to communicate with the central computer on the Oahu Island without using phone lines.

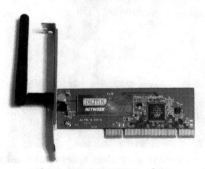

54 Mbit/s WLAN PCI Card (802.11g)

WLAN (Wireless Local Area Network) hardware initially cost so much that it was only used as an alternative to cabled LAN in places where cabling was difficult or impossible. Early development included industry-specific solutions and proprietary protocols, but at the end of the 1990s these were replaced by standards, primarily the various versions of IEEE 802.11 (in products using the Wi-Fi brand name). Beginning in 1991, a European alternative known as HiperLAN/1 was pursued by the European Telecommunications Standards Institute (ETSI) with a first version approved in 1996. This was followed by a HiperLAN/2 functional specification with ATM influences accomplished February 2000. Neither European standard achieved the commercial success of 802.11, although much of the work on HiperLAN/2 has survived in the PHY specification for IEEE 802.11a, which is nearly identical to the PHY of HiperLAN/2.

In 2009 802.11n was added to 802.11. It operates in both the 2.4 GHz and 5 GHz bands at a maximum data transfer rate of 600 Mbit/s. Most newer routers are able to utilise both wireless bands, known as dualband. This allows data communications to avoid the crowded 2.4 GHz band, which is also shared with Bluetooth devices and microwave ovens. The 5 GHz band is also wider than the 2.4 GHz band, with more channels, which permits a greater number of devices to share the space. Not all channels are available in all regions.

A HomeRF group formed in 1997 to promote a technology aimed for residential use, but it disbanded at the end of 2002.

Architecture

Stations

All components that can connect into a wireless medium in a network are referred to as stations (STA). All stations are equipped with wireless network interface controllers (WNICs). Wireless stations fall into two categories: wireless access points, and clients. Access points (APs), normally wireless routers, are base stations for the wireless network. They transmit and receive radio frequencies for wireless enabled devices to communicate with. Wireless clients can be mobile devices such as laptops, personal digital assistants, IP phones and other smartphones, or non-portable devices such as desktop computers and workstations that are equipped with a wireless network interface.

Basic Service Set

The basic service set (BSS) is a set of all stations that can communicate with each other at PHY layer. Every BSS has an identification (ID) called the BSSID, which is the MAC address of the access point servicing the BSS.

There are two types of BSS: Independent BSS (also referred to as IBSS), and infrastructure BSS. An independent BSS (IBSS) is an ad hoc network that contains no access points, which means they cannot connect to any other basic service set.

Extended Service Set

An extended service set (ESS) is a set of connected BSSs. Access points in an ESS are connected by a distribution system. Each ESS has an ID called the SSID which is a 32-byte (maximum) character string.

Distribution System

A distribution system (DS) connects access points in an extended service set. The concept of a DS can be used to increase network coverage through roaming between cells.

DS can be wired or wireless. Current wireless distribution systems are mostly based on WDS or MESH protocols, though other systems are in use.

Types of Wireless LANs

The IEEE 802.11 has two basic modes of operation: infrastructure and *ad hoc* mode. In *ad hoc* mode, mobile units transmit directly peer-to-peer. In infrastructure mode, mobile units communicate through an access point that serves as a bridge to other networks (such as Internet or LAN).

Since wireless communication uses a more open medium for communication in comparison to wired LANs, the 802.11 designers also included encryption mechanisms: Wired Equivalent Privacy (WEP, now insecure), Wi-Fi Protected Access (WPA, WPA2), to secure wireless computer networks. Many access points will also offer Wi-Fi Protected Setup, a quick (but now insecure) method of joining a new device to an encrypted network.

Infrastructure

Most Wi-Fi networks are deployed in infrastructure mode.

In infrastructure mode, a base station acts as a wireless access point hub, and nodes communicate through the hub. The hub usually, but not always, has a wired or fiber network connection, and may have permanent wireless connections to other nodes.

Wireless access points are usually fixed, and provide service to their client nodes within range.

Wireless clients, such as laptops, smartphones etc. connect to the access point to join the network.

Sometimes a network will have a multiple access points, with the same 'SSID' and security arrangement. In that case connecting to any access point on that network joins the client to the network. In that case, the client software will try to choose the access point to try to give the best service, such as the access point with the strongest signal.

Peer-to-peer

Peer-to-Peer or ad hoc wireless LAN

An ad hoc network (not the same as a WiFi Direct network) is a network where stations communicate only peer to peer (P2P). There is no base and no one gives permission to talk. This is accomplished using the Independent Basic Service Set (IBSS).

A WiFi Direct network is another type of network where stations communicate peer to peer.

In a Wi-Fi P2P group, the group owner operates as an access point and all other devices are clients. There are two main methods to establish a group owner in the Wi-Fi Direct group. In one approach, the user sets up a P2P group owner manually. This method is also known as Autonomous Group Owner (autonomous GO). In the second method, also called negotiation-based group creation, two devices compete based on the group owner intent value. The device with higher intent value becomes a group owner and the second device becomes a client. Group owner intent value can depend on whether the wireless device performs a cross-connection between an infrastructure WLAN service and a P2P group, remaining power in the wireless device, whether the wireless device is already a group owner in another group and/ or a received signal strength of the first wireless device.

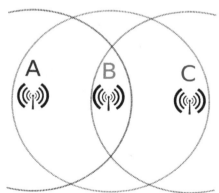

Hidden node problem: Devices A and C are both communicating with B, but are unaware of each other

A peer-to-peer network allows wireless devices to directly communicate with each other. Wireless devices within range of each other can discover and communicate directly without involving central access points. This method is typically used by two computers so that they can connect to each other to form a network. This can basically occur in devices within a closed range.

If a signal strength meter is used in this situation, it may not read the strength accurately and can be misleading, because it registers the strength of the strongest signal, which may be the closest computer.

IEEE 802.11 defines the physical layer (PHY) and MAC (Media Access Control) layers based on CSMA/CA (Carrier Sense Multiple Access with Collision Avoidance). The 802.11 specification includes provisions designed to minimize collisions, because two mobile units may both be in range of a common access point, but out of range of each other.

Bridge

A bridge can be used to connect networks, typically of different types. A wireless Ethernet bridge allows the connection of devices on a wired Ethernet network to a wireless network. The bridge acts as the connection point to the Wireless LAN.

Wireless Distribution System

A Wireless Distribution System enables the wireless interconnection of access points in an IEEE 802.11 network. It allows a wireless network to be expanded using multiple access points without the need for a wired backbone to link them, as is traditionally required. The notable advantage of WDS over other solutions is that it preserves the MAC addresses of client packets across links between access points.

An access point can be either a main, relay or remote base station. A main base station is typically connected to the wired Ethernet. A relay base station relays data between remote base stations, wireless clients or other relay stations to either a main or another relay base station. A remote base station accepts connections from wireless clients and passes them to relay or main stations. Connections between "clients" are made using MAC addresses rather than by specifying IP assignments.

All base stations in a Wireless Distribution System must be configured to use the same radio channel, and share WEP keys or WPA keys if they are used. They can be configured to different service set identifiers. WDS also requires that every base station be configured to forward to others in the system as mentioned above.

WDS may also be referred to as repeater mode because it appears to bridge and accept wireless clients at the same time (unlike traditional bridging). It should be noted, however, that throughput in this method is halved for all clients connected wirelessly.

When it is difficult to connect all of the access points in a network by wires, it is also possible to put up access points as repeaters.

Roaming

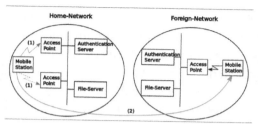

Roaming among Wireless Local Area Networks

There are two definitions for wireless LAN roaming:

1. Internal Roaming: The Mobile Station (MS) moves from one access point (AP) to another AP within a home network because the signal strength is too weak. An authentication server (RADIUS) performs the re-authentication of MS via 802.1x (e.g. with PEAP). The billing of QoS is in the home network. A Mobile Station roaming from one access point to another often interrupts the flow of data among the Mobile Station and an application connected to the network. The Mobile Station, for instance, periodically monitors the presence of alternative access points (ones that will provide a better connection). At some point, based on proprietary mechanisms, the Mobile Station decides to re-associate with an access point having a stronger wireless signal. The Mobile Station, however, may lose a connection with an access point before associating with another access point. In order to provide reliable connections with applications, the Mobile Station must generally include software that provides session persistence.

2. External Roaming: The MS (client) moves into a WLAN of another Wireless Internet Service Provider (WISP) and takes their services (Hotspot). The user can independently of his home network use another foreign network, if this is open for visitors. There must be special authentication and billing systems for mobile services in a foreign network.

Applications

Wireless LANs have a great deal of applications. Modern implementations of WLANs range from small in-home networks to large, campus-sized ones to completely mobile networks on airplanes and trains.

Users can access the Internet from WLAN hotspots in restaurants, hotels, and now with portable devices that connect to 3G or 4G networks. Oftentimes these types of public access points require no registration or password to join the network. Others can be accessed once registration has occurred and/or a fee is paid.

Existing Wireless LAN infrastructures can also be used to work as indoor positioning systems with no modification to the existing hardware.

Performance and Throughput

WLAN, organised in various layer 2 variants (IEEE 802.11), has different characteristics. Across all flavours of 802.11, maximum achievable throughputs are either given based on measurements under ideal conditions or in the layer 2 data rates. This, however, does not apply to typical deployments in which data are being transferred between two endpoints of which at least one is typically connected to a wired infrastructure and the other endpoint is connected to an infrastructure via a wireless link.

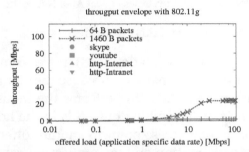

Graphical representation of Wi-Fi application specific (UDP) performance envelope 2.4 GHz band, with 802.11g

This means that typically data frames pass an 802.11 (WLAN) medium and are being converted to 802.3 (Ethernet) or vice versa.

Due to the difference in the frame (header) lengths of these two media, the packet size of an application determines the speed of the data transfer. This means that an application which uses small packets (e.g. VoIP) creates a data flow with a high overhead traffic (e.g. a low goodput).

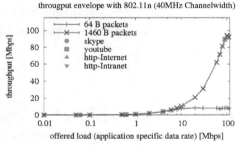

Graphical representation of Wi-Fi application specific (UDP) performance envelope 2.4 GHz band, with 802.11n with 40 MHz

Other factors which contribute to the overall application data rate are the speed with which the application transmits the packets (i.e. the data rate) and the energy with which the wireless signal is received.

The latter is determined by distance and by the configured output power of the communicating devices.

Same references apply to the attached throughput graphs which show measurements of UDP throughput measurements. Each represents an average (UDP) throughput (the error bars are there, but barely visible due to the small variation) of 25 measurements.

Each is with a specific packet size (small or large) and with a specific data rate (10 kbit/s – 100 Mbit/s). Markers for traffic profiles of common applications are included as well. This text and measurements do not cover packet errors but information about this can be found at above references. The table below shows the maximum achievable (application specific) UDP throughput in the same scenarios (same references again) with various difference WLAN (802.11) flavours. The measurement hosts have been 25 meters apart from each other; loss is again ignored.

References

- Pommer, Hermann: Roaming zwischen Wireless Local Area Networks. VDM Verlag, Saarbrücken 2008, ISBN 978-3-8364-8708-5.

- "Towards Energy-Awareness in Managing Wireless LAN Applications". IEEE/IFIP NOMS 2012: IEEE/IFIP Network Operations and Management Symposium. Retrieved 2014-08-11.

- "Application Level Energy and Performance Measurements in a Wireless LAN". The 2011 IEEE/ACM International Conference on Green Computing and Communications. Retrieved 2014-08-11.

- Marsan, Carolyn Duffy (25 June 2001). "Starbucks wireless network a sweet deal for MobileStar". Retrieved 13 April 2013.

- "American Airlines and MobileStar Network to Deliver Wireless Internet Connectivity to American's Passengers". PR Newswire. 11 May 2000. Retrieved 13 April 2013.

- "MobileStar Network to Supply U.S. Hilton Hotels With Wireless High-Speed Internet Access". 28 October 1998. Retrieved 13 April 2013.

- Brownlee, John (2013-06-12). "iOS 7 Will Make It Possible To Roam Between Open Wi-Fi Networks Without Your Data Ever Dropping". Cult of Mac. Retrieved 2013-09-16.

- Branscombe, Mary (3 October 2012). "Wi-Fi roaming: Hotspot 2.0 and Next Generation Hotspot". ZDNet. Retrieved 13 April 2013.

- "Internet Security Podcast episode 10: Free WiFi And The Security issues it poses". 18 February 2013. Retrieved 13 April 2013.

- vonNagy, Andrew (20 May 2012). "Wi-Fi Alliance Rebrands Hotspot 2.0 as Wi-Fi Certified Passpoint". Retrieved 13 April 2012.

- Vos, Esme. "Picocela Deploys Large Mesh Wifi Hotzone in Fukuoka Japan". Muniwireless Blog. Muniwireless.com. Retrieved 8 April 2011.

6

Related Aspects of WiMAX Technology

Several new services have emerged that utilize WiMAX like WiBro, cognitive radio, category 5 cable, high speed packet access, last mile, wireless local loop, customer-premises equipment etc. The chapter discusses the difference that WiMAX has brought to these services and how these services are revolutionizing the transmission of wireless communication. The aspects elucidated in this chapter are of vital importance, and provide a better understanding of WiMAX technology

WiBro

WiBro (Wireless Broadband) is a wireless broadband Internet technology developed by the South Korean telecoms industry. WiBro is the South Korean service name for IEEE 802.16e (mobile WiMAX) international standard. By the end of 2012, the Korean Communications Commission intends to increase WiBro broadband connection speeds to 10Mbit/s, around ten times the current speed, which will complement their 1Gbit/sec fibre-optic network.

wibro modem KWD-B2800(strong egg sold by kt, produced by modacom)

WiBro adopts TDD for duplexing, OFDMA for multiple access and 8.75/10.00 MHz as a channel bandwidth. WiBro was devised to overcome the data rate limitation of mobile phones (for example CDMA 1x) and to add mobility to broadband Internet access (for example ADSL or Wireless LAN). In February 2002, the Korean government allocated 100 MHz of electromagnetic spectrum in the 2.3–2.4 GHz band, and in late 2004 WiBro Phase 1 was standardized by the TTA of Korea and in late 2005 ITU reflected WiBro as IEEE 802.16e (mobile WiMAX). Two South Korean Telcom (KT, SKT) launched commercial service in June 2006, and the tariff is around US$30.

WiBro base stations offer an aggregate data throughput of 30 to 50 Mbit/s per carrier and cover a radius of 1–5 km allowing for the use of portable internet usage. In detail, it provides mobility for moving devices up to 120 km/h (74.5 mi/h) compared to Wireless LAN having mobility up to walking speed and mobile phone technologies having mobility up to 250 km/h. From testing during the APEC Summit in Busan in late 2005, the actual range and bandwidth were quite a bit lower than these numbers. The technology will also offer Quality of Service. The inclusion of QoS allows for WiBro to stream video content and other loss-sensitive data in a reliable manner. These all appear to be (and may be) the stronger advantages over the fixed WiMAX standard (802.16a). Some Telcos in many countries were trying to commercialize this Mobile WiMAX (or WiBro). For example, TI (Italy), TVA (Brazil), Omnivision (Venezuela), PORTUS (Croatia), and Arialink (Michigan) provided commercial service at some stage. While WiBro is quite precise in its requirements from spectrum use to equipment design, WiMAX leaves much of this up to the equipment provider while providing enough detail to ensure interoperability between designs.

WiBro has a Peak Download speed of 128 Mbit/s and a Peak Upload speed of 56 Mbit/s.

Current Service

In Korea, KT (Korea Telecom) offers Wave 2 (18.4 Mbit/s, 4 Mbit/s) for 10000 KRW/mo (around $11 or €6.50) with 10 GB data usage to 40000 KRW/mo with 50 GB data usage plus free access to their own WiFi hotspots, ollehWiFi. The service coverage is advertised as nationwide, but actual coverage is restricted to every city, some railroad station, airports, and major highways. SK Telecom also offers Wave 2 WiBro Service for $18.87 a month with 30 GB data usage. Actual service coverage is limited mostly to major cities and highways.

For short term visitors, KT rents WiBro modem and bridge at KT Roaming Center in Incheon International Airport. As of October 2012, WiBro-only USB modem costs 5,000 KRW per day plus 100,000 KRW deposit, WiBro-HSPA USB modem and WiBro Egg cost 8,000 KRW per day plus 150,000 KRW deposit. One-day rent is free at KT Roaming Center, and requires credit card and passport.

In India, Tikona Digital Networks (Independent services provider) offers WiBro service for up to 2 Mbit/s and 4 Mbit/s in many cities. The 2 MBit/s unlimited monthly plan costs Rs. 999.00 (roughly $21).

Coverage

As of January 2013, KT covers all 80+ cities while SK covers Seoul and a few other major cities in South Korea.

Wibro Manufactures

- HTC - HTC Evo 4G+

- INFOMARK - Compact Egg (Mobile router)

- Interbro - Egg (A mobile Wibro to Wi-Fi hotspot device brand)

- Intel - Wibro Netbook

- LG Innotek - Egg, USB Type Modem

- Modacom - Egg

- Myungmin - USB Type Modem

- Samsung - USB Type Modem, Wibro 3G phone (SCH-M830, Show Wibro Omnia)

- Iphone4 - USIM

Network Deployment

In November 2004, Intel and Samsung Electronics executives agreed to ensure compatibility between WiBro and Mobile WiMAX technology.

KT Corporation, SK Telecom and Hanaro Telecom (acquired by SK Telecom and renamed SK Broadband) had been selected as Wibro operators in January, 2005. However, Hanaro Telecom cancelled its plan for the WiBro and returned WiBro licence in April 2005.

In September 2005, Samsung Electronics signed a deal with Sprint Nextel Corporation to provide equipment for a WiBro trial.

In November 2005, KT Corporation (aka Korea Telecom) showed off WiBro trial services during the Asia-Pacific Economic Cooperation (APEC) summit in Busan.

February 10, 2006: Telecom Italia, the dominant telephony and internet service provider in Italy, together with Korean Samsung Electronics, has demonstrated to the public a WiBro network service on the occasion of the 2006 Winter Olympics, held in Turin, with downlink speed of 10 Mbit/s and uplink speed of some hundreds of kbit/s even in movement up to 120 km/h.

In the same event Samsung tlc div. president Kitae Lee assured a future of 20–30 Mbit/s by the end of this year (2006) and >100 Mbit/s down/>1 Mbit/s up in 2008.

KT Corporation launched commercial WiBro service on June 2006.

Sprint (US), BT (UK), KDDI (JP), and TVA (BR) have or are trialing WiBro.

KT Corporation and SK Telecom launched WiBro around Seoul on June 30, 2006. More about the KT launch.

On April 3, 2007, KT launched WiBro coverage for all areas of Seoul including all subway lines.

In January 2011, KT's mobile network SHOW and home network QOOK merged. Since then, KT has been changed to olleh.

In March, 2011, olleh's WiBro coverage was expanded nationwide covering 85% of Koreans.

As of October 2012, olleh's Wibro covers 88% of the South Korean population.

Cognitive Radio

A cognitive radio (CR) is an intelligent radio that can be programmed and configured dynamically. Its transceiver is designed to use the best wireless channels in its vicinity. Such a radio automatically detects available channels in wireless spectrum, then accordingly changes its transmission or reception parameters to allow more concurrent wireless communications in a given spectrum band at one location. This process is a form of dynamic spectrum management.

Description

In response to the operator's commands, the cognitive engine is capable of configuring radio-system parameters. These parameters include "waveform, protocol, operating frequency, and networking". This functions as an autonomous unit in the communications environment, exchanging information about the environment with the networks it accesses and other cognitive radios (CRs). A CR "monitors its own performance continuously", in addition to "reading the radio's outputs"; it then uses this information to "determine the RF environment, channel conditions, link performance, etc.", and adjusts the "radio's settings to deliver the required quality of service subject to an appropriate combination of user requirements, operational limitations, and regulatory constraints".

Some "smart radio" proposals combine wireless mesh network—dynamically changing the path messages take between two given nodes using cooperative diversity; cognitive radio—dynamically changing the frequency band used by messages between two consecutive nodes on the path; and software-defined radio—dynamically changing the protocol used by message between two consecutive nodes.

J. H. Snider, Lawrence Lessig, David Weinberger, and others say that low power "smart" radio is inherently superior to standard broadcast radio.

History

The concept of cognitive radio was first proposed by Joseph Mitola III in a seminar at KTH (the Royal Institute of Technology in Stockholm) in 1998 and published in an article by Mitola and Gerald Q. Maguire, Jr. in 1999. It was a novel approach in wireless communications, which Mitola later described as:

The point in which wireless personal digital assistants (PDAs) and the related networks are sufficiently computationally intelligent about radio resources and related computer-to-computer communications to detect user communications needs as a function of use context, and to provide radio resources and wireless services most appropriate to those needs.

Cognitive radio is considered as a goal towards which a software-defined radio platform should evolve: a fully reconfigurable wireless transceiver which automatically adapts its communication parameters to network and user demands.

Traditional regulatory structures have been built for an analog model and are not optimized for cognitive radio. Regulatory bodies in the world (including the Federal Communications Commission in the United States and Ofcom in the United Kingdom) as well as different independent measurement campaigns found that most radio frequency spectrum was inefficiently utilized. Cellular network bands are overloaded in most parts of the world, but other frequency bands (such as military, amateur radio and paging frequencies) are insufficiently utilized. Independent studies performed in some countries confirmed that observation, and concluded that spectrum utilization depends on time and place. Moreover, fixed spectrum allocation prevents rarely used frequencies (those assigned to specific services) from being used, even when any unlicensed users would not cause noticeable interference to the assigned service. Regulatory bodies in the world have been considering whether to allow unlicensed users in licensed bands if they would not cause any interference to licensed users. These initiatives have focused cognitive-radio research on dynamic spectrum access.

Terminology

Depending on transmission and reception parameters, there are two main types of cognitive radio:

- *Full Cognitive Radio* (Mitola radio), in which every possible parameter observable by a wireless node (or network) is considered.

- *Spectrum-Sensing Cognitive Radio*, in which only the radio-frequency spectrum is considered.

Other types are dependent on parts of the spectrum available for cognitive radio:

- *Licensed-Band Cognitive Radio*, capable of using bands assigned to licensed users (except for unlicensed bands, such as the U-NII band or the ISM band. The IEEE 802.22 working group is developing a standard for wireless regional area network (WRAN), which will operate on unused television channels.

- *Unlicensed-Band Cognitive Radio*, which can only utilize unlicensed parts of the radio frequency (RF) spectrum. One such system is described in the IEEE 802.15 Task Group 2 specifications, which focus on the coexistence of IEEE 802.11 and Bluetooth.

- *Spectrum mobility*: Process by which a cognitive-radio user changes its frequency of operation. Cognitive-radio networks aim to use the spectrum in a dynamic manner by allowing radio terminals to operate in the best available frequency band, maintaining seamless communication requirements during transitions to better spectrum.

- *Spectrum sharing*: Spectrum sharing cognitive radio networks allow cognitive radio users to share the spectrum bands of the licensed-band users. However, the cognitive radio users have to restrict their transmit power so that the interference caused to the licensed-band users is kept below a certain threshold.

- *Sensing-based Spectrum sharing*: In sensing-based spectrum sharing cognitive radio networks, cognitive radio users first listen to the spectrum allocated to the licensed users to detect the state of the licensed users. Based on the detection results, cognitive radio users decide their transmission strategies. If the licensed users are not using the bands, cognitive radio users will transmit over those bands. If the licensed users are using the bands, cognitive radio users share the spectrum bands with the licensed users by restricting their transmit power.

Technology

Although cognitive radio was initially thought of as a software-defined radio extension (full cognitive radio), most research work focuses on spectrum-sensing cognitive radio (particularly in the TV bands). The chief problem in spectrum-sensing cognitive radio is designing high-quality spectrum-sensing devices and algorithms for exchanging spectrum-sensing data between nodes. It has been shown that a simple energy detector cannot guarantee the accurate detection of signal presence, calling for more sophisticated spectrum sensing techniques and requiring information about spectrum sensing to be regularly exchanged between nodes. Increasing the number of cooperating sensing nodes decreases the probability of false detection.

Filling free RF bands adaptively, using OFDMA, is a possible approach. Timo A. Weiss and Friedrich K. Jondral of the University of Karlsruhe proposed a spectrum pooling system, in which free bands (sensed by nodes) were immediately filled by OFDMA sub-bands. Applications of spectrum-sensing cognitive radio include emergency-network and WLAN higher throughput and transmission-distance extensions. The evolution of cognitive radio toward cognitive networks is underway; the concept of cognitive networks is to intelligently organize a network of cognitive radios.

Functions

The main functions of cognitive radios are:

- *Power Control*: Power control is usually used for spectrum sharing CR systems to maximize the capacity of secondary users with interference power constraints to protect the primary users.

- *Spectrum sensing*: Detecting unused spectrum and sharing it, without harmful interference to other users; an important requirement of the cognitive-radio network is to sense empty spectrum. Detecting primary users is the most efficient way to detect empty spectrum. Spectrum-sensing techniques may be grouped into three categories:

  - *Transmitter detection*: Cognitive radios must have the capability to determine if a signal from a primary transmitter is locally present in a certain spectrum. There are several proposed approaches to transmitter detection:

- Matched filter detection

- Energy detection: Energy detection is a spectrum sensing method that detects the presence/absence of a signal just by measuring the received signal power. This signal detection approach is quite easy and convenient for practical implementation. To implement energy detector, however, noise variance information is required. It has been shown that an imperfect knowledge of the noise power (noise uncertainty) may lead to the phenomenon of the SNR wall, which is a SNR level below which the energy detector can not reliably detect any transmitted signal even increasing the observation time. It has also been shown that the SNR wall is not caused by the presence of a noise uncertainty itself, but by an insufficient refinement of the noise power estimation while the observation time increases.

- Cyclostationary-feature detection: These type of spectrum sensing algorithms are motivated because most man-made communication signals, such as BPSK, QPSK, AM, OFDM, etc. exhibit cyclostationary behavior. However, noise signals (typically white noise) do not exhibit cyclostationary behavior. These detectors are robust against noise variance uncertainty. The aim of such detectors is to exploit the cyclostationary nature of man-made communication signals buried in noise. Cyclostationary detectors can be either single cycle or multicycle cyclostatonary.

- *Wideband spectrum sensing*: refers to spectrum sensing over large spectral bandwidth, typically hundreds of MHz or even several GHz. Since current ADC technology cannot afford the high sampling rate with high resolution, it requires revolutional techniques, e.g., compressive sensing and sub-Nyquist sampling.

- ○ *Cooperative detection*: Refers to spectrum-sensing methods where information from multiple cognitive-radio users is incorporated for primary-user detection

- ○ *Interference-based detection*

- *Null-space based CR*: With the aid of multiple antennas, CR detects the null-space of the primary-user and then transmits within the null-space, such that its subsequent transmission causes less interference to the primary-user

- *Spectrum management*: Capturing the best available spectrum to meet user communication requirements, while not creating undue interference to other (primary) users. Cognitive radios should decide on the best spectrum band (of all bands available) to meet quality of service requirements; therefore, spectrum-management functions are required for cognitive radios. Spectrum-management functions are classified as:

  - ○ *Spectrum analysis*

  - ○ *Spectrum decision*

The practical implementation of spectrum-management functions is a complex and multifaceted issue, since it must address a variety of technical and legal requirements. An example of the former is choosing an appropriate sensing threshold to detect other users, while the latter is exemplified by the need to meet the rules and regulations set out for radio spectrum access in international (ITU radio regulations) and national (telecommunications law) legislation.

Versus Intelligent Antenna (IA)

An intelligent antenna (or smart antenna) is an antenna technology that uses spatial beam-formation and spatial coding to cancel interference; however, applications are emerging for extension to intelligent multiple or cooperative-antenna arrays for application to complex communication environments. Cognitive radio, by comparison, allows user terminals to sense whether a portion of the spectrum is being used in order to share spectrum with neighbor users. The following table compares the two:

Note that both techniques can be combined as illustrated in many contemporary transmission scenarios.

Cooperative MIMO (CO-MIMO) combines both techniques.

Applications

CR can sense its environment and, without the intervention of the user, can adapt to the user's communications needs while conforming to FCC rules in the United States. In theory, the amount of spectrum is infinite; practically, for propagation and other reasons it is finite because of the desirability of certain spectrum portions. Assigned spectrum is far from being fully utilized, and efficient spectrum use is a growing concern; CR offers a solution to this problem. A CR can intelligently detect whether any portion of the spectrum is in use, and can temporarily use it without interfering with the transmissions of other users. According to Bruce Fette, "Some of the radio's other cognitive abilities include determining its location, sensing spectrum use by neighboring devices, changing frequency, adjusting output power or even altering transmission parameters and characteristics. All of these capabilities, and others yet to be realized, will provide wireless spectrum users with the ability to adapt to real-time spectrum conditions, offering regulators, licenses and the general public flexible, efficient and comprehensive use of the spectrum".

Simulation of CR Networks

At present, modeling & simulation is the only paradigm which allows the simulation of complex behavior in a given environment's cognitive radio networks. Network simulators like OPNET, NetSim, MATLAB and NS2 can be used to simulate a cognitive radio network. Areas of research using network simulators include:

1. Spectrum sensing & incumbent detection

2. Spectrum allocation

3. Measurement and modeling of spectrum usage

4. Efficiency of spectrum utilization

Future Plans

The success of the unlicensed band in accommodating a range of wireless devices and services has led the FCC to consider opening further bands for unlicensed use. In contrast, the licensed bands are underutilized due to static frequency allocation. Realizing that CR technology has the potential to exploit the inefficiently utilized licensed bands without causing interference to incumbent users, the FCC released a Notice of Proposed Rule Making which would allow unlicensed radios to operate in the TV-broadcast bands. The IEEE 802.22 working group, formed in November 2004, is tasked with defining the air-interface standard for wireless regional area networks (based on CR sensing) for the operation of unlicensed devices in the spectrum allocated to TV service.

Category 5 Cable

Category 5 cable, commonly referred to as cat 5, is a twisted pair cable for carrying signals. This type of cable is used in structured cabling for computer networks such as

Ethernet. The cable standard provides performance of up to 100 MHz and is suitable for 10BASE-T, 100BASE-TX (Fast Ethernet), and 1000BASE-T (Gigabit Ethernet). Cat 5 is also used to carry other signals such as telephony and video.

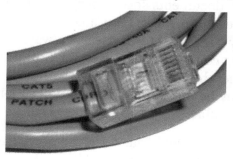

Category 5 patch cable in T568B wiring

This cable is commonly connected using punch-down blocks and modular connectors. Most category 5 cables are unshielded, relying on the balanced line twisted pair design and differential signaling for noise rejection.

Category 5 was superseded by the category 5e (enhanced) specification, and later category 6 cable.

Partially stripped cable showing the twisted pairs.

A cat 5e Wall outlet showing the two wiring schemes: A for T568A, B for T568B.

Cable Standard

The specification for category 5 cable was defined in ANSI/TIA/EIA-568-A, with clarification in TSB-95. These documents specify performance characteristics and test requirements for frequencies up to 100 MHz. Cable types, connector types and cabling topologies are defined by TIA/EIA-568-B. Nearly always, 8P8C modular connectors (often referred to as RJ45 connectors) are used for connecting category 5 cable. The cable is terminated in either the T568A scheme or the T568B scheme. The two schemes work equally well and may be mixed in an installation so long as the same scheme is used on both ends of each cable.

Each of the four pairs in a cat 5 cable has differing precise number of twists per meter to minimize crosstalk between the pairs. Although cable assemblies containing 4 pairs are common, category 5 is not limited to 4 pairs. Backbone applications involve using up to 100 pairs. This use of balanced lines helps preserve a high signal-to-noise ratio despite interference from both external sources and crosstalk from other pairs.

The cable is available in both stranded and solid conductor forms. The stranded form is more flexible and withstands more bending without breaking. Permanent wiring (for example, the wiring inside the wall that connects a wall socket to a central patch panel) is solid-core, while patch cables (for example, the movable cable that plugs into the wall socket on one end and a computer on the other) are stranded.

The specific category of cable in use can be identified by the printing on the side of the cable.

Bending Radius

Most Category 5 cables can be bent at any radius exceeding approximately four times the outside diameter of the cable.

Maximum Cable Segment Length

The maximum length for a cable segment is 100 m per TIA/EIA 568-5-A. If longer runs are required, the use of active hardware such as a repeater or switch is necessary. The specifications for 10BASE-T networking specify a 100-meter length between active devices. This allows for 90 meters of solid-core permanent wiring, two connectors and two stranded patch cables of 5 meters, one at each end.

Category 5 vs. 5e

The category 5e specification improves upon the category 5 specification by tightening some crosstalk specifications and introducing new crosstalk specifications that were not present in the original category 5 specification. The bandwidth of category 5 and 5e is the same (100 MHz) and the physical cable construction is the same, and the reality is that most Cat 5 cables meet Cat 5e specifications, though it is not tested or certified as such.

Category 5e Vs. 6

The category 6 specification improves upon the category 5e specification by improving frequency response, tightening crosstalk specifications, and introducing more comprehensive crosstalk specifications. The improved performance of Cat 6 is 250 MHz and supports 10GBASE-T (10-Gigabit Ethernet).

Applications

This type of cable is used in structured cabling for computer networks such as Ethernet over twisted pair. The cable standard provides performance of up to 100 MHz and is suitable for 10BASE-T, 100BASE-TX (Fast Ethernet), and 1000BASE-T (Gigabit Ethernet). 10BASE-T and 100BASE-TX Ethernet connections require two wire pairs. 1000BASE-T Ethernet connections require four wire pairs. Through the use of power over Ethernet (PoE), up to 25 watts of power can be carried over the cable in addition to Ethernet data.

Cat 5 is also used to carry other signals such as telephony and video.

Shared Cable

In some cases, multiple signals can be carried on a single cable; cat 5 can carry two conventional telephone lines as well as 100BASE-TX in a single cable. The USOC/RJ-61 wiring standard may be used in multi-line telephone connections.

Various schemes exist for transporting both analog and digital video over the cable. HDBaseT (10.2 Gbit/s) is one such scheme.

Characteristics

Insulation

Outer insulation is typically polyvinyl chloride (PVC) or low smoke zero halogen (LSOH).

Conductors

Since 1995, solid-conductor UTP cables for backbone cabling is required to be no thicker than 22 American Wire Gauge (AWG) and no thinner than 24 AWG, or 26 AWG for shorter-distance cabling. This standard has been retained with the 2009 revision of ANSI TIA/EIA 568.

Individual Twist Lengths

By altering the length of each twist, crosstalk is reduced, without affecting the characteristic impedance. The distance per twist is commonly referred to as pitch. The pitch of the twisted pairs is not specified in the standard. Measurements on one sample of cat 5 cable yielded the following results. Since the pitch of the various colors is not specified in the standard, pitch can vary according to manufacturer and should be measured for the batch being used if cable is being used in non-Ethernet situation where pitch might be critical.

Environmental Ratings

> Communications riser (CMR) is insulated with high-density polyolefin and jacketed with low-smoke polyvinyl chloride (PVC).

> Communications plenum (CMP) is insulated with fluorinated ethylene propylene (FEP) and polyethylene (PE) and jacketed with low-smoke polyvinyl chloride (PVC), due to better flame test ratings.

> Communications (CM) is insulated with high-density polyolefin, but not jacketed with PVC and therefore is the lowest of the three in flame resistance.

Some cables are "UV-rated" or "UV-stable" meaning they can be exposed to outdoor UV radiation without significant destruction.

Plenum-rated cables are slower to burn and produce less smoke than cables using a mantle of materials like PVC. This also affects legal requirements for a fire sprinkler system. That is if a plenum-rated cable is used, sprinkler requirement may be eliminated.

Shielded cables (FTP or STP) are useful for environments where proximity to RF equipment may introduce electromagnetic interference, and can also be used where eavesdropping likelihood should be minimized.

Mobile Broadband

A mobile broadband modem in the ExpressCard form factor for laptop computers

HTC ThunderBolt, the second commercially available LTE smartphone

Mobile broadband is the marketing term for wireless Internet access through a portable modem, mobile phone, USB wireless modem, tablet or other mobile devices. The first wireless Internet access became available in 1991 as part of the second generation (2G) of mobile phone technology. Higher speeds became available in 2001 and 2006 as part of the third (3G) and fourth (4G) generations. In 2011, 90% of the world's population lived in areas with 2G coverage, while 45% lived in areas with 2G and 3G coverage. Mobile broadband uses the spectrum of 225 MHz to 3700 MHz.

Description

Mobile broadband is the marketing term for wireless Internet access delivered through mobile phone towers to computers, mobile phones (called "cell phones" in North America and South Africa), and other digital devices using portable modems. Although broadband has a technical meaning, wireless-carrier marketing uses the phrase "mobile broadband" as a synonym for mobile Internet access. Some mobile services allow more than one device to be connected to the Internet using a single cellular connection using a process called tethering.

The bit rates available with Mobile broadband devices support voice and video as well as other data access. Devices that provide mobile broadband to mobile computers include:

- PC cards, also known as *PC data cards*, and Express cards

- USB and mobile broadband modems, also known as *connect cards*

- portable devices with built-in support for mobile broadband, such as laptop computers, netbook computers, smartphones, tablets, PDAs, and other mobile Internet devices.

Internet access subscriptions are usually sold separately from mobile phone subscriptions.

Generations

Roughly every ten years new mobile phone technology and infrastructure involving a change in the fundamental nature of the service, non-backwards-compatible transmission technology, higher peak data rates, new frequency bands, wider channel frequency bandwidth in Hertz becomes available. These transitions are referred to as generations. The first mobile data services became available during the second generation (2G).

The download (to the user) and upload (to the Internet) data rates given above are peak or maximum rates and end users will typically experience lower data rates.

WiMAX was originally developed to deliver fixed wireless service with wireless mobility added in 2005. CDPD, CDMA2000 EV-DO, and MBWA are no longer being actively developed.

Coverage

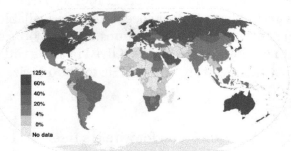

Mobile broadband Internet subscriptions in 2012 as a percentage of a country's population

Source: International Telecommunications Union.

In 2011, 90% of the world's population lived in areas with 2G coverage, while 45% lived in areas with 2G and 3G coverage, and 5% lived in areas with 4G coverage. By 2017 more than 90% of the world's population is expected to have 2G coverage, 85% is expected to have 3G coverage, and 50% will have 4G coverage.

A barrier to mobile broadband use is the coverage provided by the mobile phone networks. This may mean no mobile phone service or that service is limited to older and slower mobile broadband technologies. Customers will not always be able to achieve the speeds advertised due to mobile data coverage limitations including distance to the cell tower. In addition, there are issues with connectivity, network capacity, application quality, and mobile network operators' overall inexperience with data traffic. Peak speeds experienced by users are also often limited by the capabilities of their smartphone or other mobile device.

Subscriptions and Usage

It is estimated that there were 6.6 billion mobile phone subscriptions worldwide at the end of 2012 (89% penetration), representing roughly 4.4 billion subscribers (many people have more than one subscription). Growth has been around 9% year-on-year. Mobile phone subscriptions are expected to reach 9.3 billion in 2018.

At the end of 2012 there were roughly 1.5 billion mobile broadband subscriptions growing at a 50% year-on-year rate. Mobile broadband subscriptions are expected to reach 6.5 billion in 2018.

Mobile data traffic doubled between the end of 2011 (~620 Petabytes in Q4 2011) and the end of 2012 (~1280 Petabytes in Q4 2012). This traffic growth is and will continue to be driven by large increases in the number of mobile subscriptions and by increases in the average data traffic per subscription due to increases in the number of smartphones being sold, the use of more demanding applications and in particular video, and the availability and deployment of newer 3G and 4G technologies capable of higher data rates. By 2018 total mobile broadband traffic is expected to increase by a factor of 12 to roughly 13,000 PetaBytes.

On average, a mobile PC generates approximately seven times more traffic than a smartphone (3 GB vs. 450 MB/month). By 2018 this ratio is likely to fall to 5 times (10 GB vs. 2 GB/month). Traffic from smartphones that tether (share the data access of one device with multiple devices) can be up to 20 times higher than that from non-tethering users and averages between 7 and 14 times higher.

Note too that there are large differences in subscriber and traffic patterns between different provider networks, regional markets, device and user types.

Demand from emerging markets has and continues to fuel growth in both mobile phone and mobile broadband subscriptions and use. Lacking a widespread fixed line infrastructure, many emerging markets leapfrog developed markets and use mobile broadband technologies to deliver high-speed internet access to the mass market.

Development

Service mark for GSMA mobile broadband

In use and under Active Development

GSM Family

In 1995 telecommunication, mobile phone, integrated-circuit, and laptop computer manufacturers formed the GSM Association to push for built-in support for mobile-broadband technology on notebook computers. The association established a service mark to identify devices that include Internet connectivity. Established in early 1998, the global Third Generation Partnership Project (3GPP) develops the evolving GSM family of standards, which includes GSM, EDGE, WCDMA, HSPA, and LTE. In 2011 these standards were the most used method to deliver mobile broadband. With the development of the 4G LTE signalling standard, download speeds could be increased to 300 Mbit/s per second within the next several years.

IEEE 802.16 (WiMAX)

The IEEE working group IEEE 802.16, produces standards adopted in products using the WiMAX trademark. The original "Fixed WiMAX" standard was released in 2001 and "Mobile WiMAX" was added in 2005. The WiMAX Forum is a non-profit organization formed to promote the adoption of WiMAX compatible products and services.

In use, but Moving to Other Protocols Going Forward

CDMA Family

High Speed Packet Access

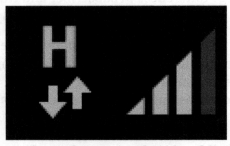

HSPA sign shown in notification bar on an Android-based (Samsung) smartphone.

High Speed Packet Access (HSPA) is an amalgamation of two mobile protocols, High Speed Downlink Packet Access (HSDPA) and High Speed Uplink Packet Access (HSUPA), that extends and improves the performance of existing 3G mobile telecommunication networks utilizing the WCDMA protocols. A further improved 3GPP standard, Evolved High Speed Packet Access (also known as HSPA+), was released late in 2008 with subsequent worldwide adoption beginning in 2010. The newer standard allows bit-rates to reach as high as 337 Mbit/s in the downlink and 34 Mbit/s in the uplink. However, these speeds are rarely achieved in practice.

Overview

The first HSPA specifications supported increased peak data rates of up to 14 Mbit/s in the downlink and 5.76 Mbit/s in the uplink. It also reduced latency and provided up to five times more system capacity in the downlink and up to twice as much system capacity in the uplink compared with original WCDMA protocol.

High Speed Downlink Packet Access (HSDPA)

High Speed Downlink Packet Access (HSDPA) is an enhanced 3G (third-generation) mobile communications protocol in the High-Speed Packet Access (HSPA) family, also dubbed 3.5G, 3G+, or Turbo 3G, which allows networks based on Universal Mobile Telecommunications System (UMTS) to have higher data speeds and capacity. HSDPA has been introduced with 3GPP Release 5, which also accompanies an improvement on the uplink providing a new bearer of 384 kbit/s. The previous maximum bearer was 128 kbit/s. As well as improving data rates, HSDPA also decreases latency and so the round trip time for applications. HSPA+ introduced in 3GPP Release 7 further increases data rates by adding 64QAM modulation, MIMO and Dual-Cell HSDPA operation, i.e. two 5 MHz carriers are used simultaneously. Even higher speeds of up to 337.5 Mbit/s are possible with Release 11 of the 3GPP standards.

The first phase of HSDPA has been specified in the 3GPP release 5. Phase one introduces new basic functions and is aimed to achieve peak data rates of 14.0 Mbit/s with significantly reduced latency. The improvement in speed and latency reduces the cost per bit and enhances support for high-performance packet data applications. HSDPA is based on shared channel transmission and its key features are shared channel and multi-code transmission, higher-order modulation, short transmission time interval (TTI), fast link adaptation and scheduling along with fast hybrid automatic repeat request (HARQ). Further new features are the High Speed Downlink Shared Channels (HS-DSCH), the adaptive modulation QPSK and 16QAM and the High Speed Medium Access protocol (MAC-hs) in base station.

The upgrade to HSDPA is often just a software update for WCDMA networks. In general voice calls are usually prioritized over data transfer.

User Equipment (UE) Categories

The following table is derived from table 5.1a of the release 11 of 3GPP TS 25.306 and shows maximum data rates of different device classes and by what combination of features they are achieved. The per-cell per-stream data rate is limited by the *Maximum number of bits of an HS-DSCH transport block received within an HS-DSCH TTI* and the *Minimum inter-TTI interval*. The TTI is 2 ms. So for example Cat 10 can decode 27952 bits/2 ms = 13.976 MBit/s (and not 14.4 MBit/s as often claimed incorrectly). Categories 1-4 and 11 have inter-TTI intervals of 2 or 3, which reduces the maximum data rate by that factor. Dual-Cell and MIMO 2x2 each multiply the maximum data rate by 2, because multiple independent transport blocks are transmitted over different carriers or spatial streams, respectively. The data rates given in the table are rounded to one decimal point.

HSDPA User Equipment (UE) categories

UE categories were defined from 3GGP Release 7 onwards as Evolved HSPA (HSPA+) and are listed in Evolved HSDPA UE Categories.

Adoption

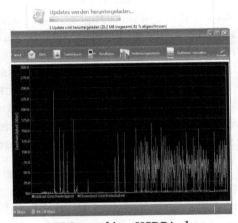

GPRS-speed in a HSDPA plan

As of 28 August 2009, 250 HSDPA networks have commercially launched mobile broadband services in 109 countries. 169 HSDPA networks support 3.6 Mbit/s peak downlink data throughput. A growing number are delivering 21 Mbit/s peak data downlink and 28 Mbit/s.

CDMA2000-EVDO networks had the early lead on performance, and Japanese providers were highly successful benchmarks for it. But lately this seems to be changing in favour of HSDPA as an increasing number of providers worldwide are adopting it.

During 2007, an increasing number of telcos worldwide began selling HSDPA USB modems to provide mobile broadband connections. In addition, the popularity of HSDPA landline replacement boxes grew—providing HSDPA for data via Ethernet and WiFi, and ports for connecting traditional landline telephones. Some are marketed with connection speeds of "up to 7.2 Mbit/s", which is only attained under ideal conditions. As a result, these services can be slower than expected, when in fringe coverage indoors.

High Speed Uplink Packet Access (HSUPA)

High-Speed Uplink Packet Access (HSUPA) is a 3G mobile telephony protocol in the

HSPA family. This technology was the second major step in the UMTS evolution process. It was specified and standardized in 3GPP Release 6 to improve the uplink data rate to 5.76 Mbit/s, extending the capacity, and reducing latency. Together with additional improvements which are detaileld below this creates opportunities for a number of new applications including VoIP, uploading pictures and sending large e-mail messages.

In the meanwhile HSUPA has been superseded by newer technologies further advancing transfer rates. LTE provides up to 300 Mbit/s for downlink and 75 Mbit/s for uplink. Its evolution LTE Advanced supports maximum downlink rates of over 1 Gbit/s.

Technology

Enhanced Uplink adds a new transport channel to WCDMA, called the Enhanced Dedicated Channel (E-DCH). Further it features several improvements similar to those of HSDPA, including multi-code transmission, shorter Transmission Time Interval (TTI) enabling faster link adaptation, fast scheduling and fast Hybrid Automatic Repeat Request (HARQ) with incremental redundancy making retransmissions more effective. Similarly to HSDPA, HSUPA uses a *packet scheduler*, but it operates on a *request-grant* principle where the UEs request a permission to send data and the scheduler decides when and how many UEs will be allowed to do so. A request for transmission contains data about the state of the transmission buffer and the queue at the UE and its available power margin. However, unlike HSDPA, uplink transmissions are not orthogonal to each other.

In addition to this *scheduled* mode of transmission the standards also allows a self-initiated transmission mode from the UEs, denoted *non-scheduled*. The *non-scheduled* mode can, for example, be used for VoIP services for which even the reduced TTI and the Node B based scheduler will not be able to provide the very short delay time and constant bandwidth required.

Each MAC-d flow (i.e. QoS flow) is configured to use either *scheduled* or *non-scheduled* modes; the UE adjusts the data rate for *scheduled* and *non-scheduled* flows independently. The maximum data rate of each *non-scheduled* flow is configured at call setup, and typically not changed frequently. The power used by the *scheduled* flows is controlled dynamically by the Node B through absolute grant (consisting of an actual value) and relative grant (consisting of a single up/down bit) messages.

At the Physical Layer, HSUPA introduces new channels E-AGCH (Absolute Grant Channel), E-RGCH (Relative Grant Channel), F-DPCH (Fractional-DPCH), E-HICH (E-DCH Hybrid ARQ Indicator Channel), E-DPCCH (E-DCH Dedicated Physical Control Channel) and E-DPDCH (E-DCH Dedicated Physical Data Channel).

E-DPDCH is used to carry the E-DCH Transport Channel; and E-DPCCH is used to carry the control information associated with the E-DCH.

User Equipment (UE) Categories

The following table shows uplink speeds for the different categories of HSUPA.

HSUPA User Equipment (UE) categories

UE categories were defined from 3GGP Release 7 onwards as Evolved HSPA (HSPA+) and are listed in Evolved HSUPA UE Categories.

Evolved High Speed Packet Access (HSPA+)

Evolved HSPA (also known as HSPA Evolution, HSPA+) is a wireless broadband standard defined in 3GPP release 7 of the WCDMA specification. It provides extensions to the existing HSPA definitions and is therefore backward-compatible all the way to the original Release 99 WCDMA network releases. Evolved HSPA provides data rates up to 168 Mbit/s in the downlink and 22 Mbit/s in the uplink (per 5 MHz carrier) with multiple input, multiple output (2x2 MIMO) technologies and higher order modulation (64 QAM). With Dual Cell technology, these can be doubled.

Since 2011, HSPA+ has been very widely deployed amongst WCDMA operators with nearly 200 commitments.

Evolved High Speed Packet Access

HSPA+ sign shown in notification bar on an Android-based smartphone.

Evolved High Speed Packet Access, or HSPA+, is a technical standard for wireless, broadband telecommunication. It is the second phase of HSPA which has been introduced in 3GPP release 7 and being further improved in later 3GPP releases. HSPA+ can achieve data rates of up to 42.2 Mbit/s. It introduces antenna array technologies such as beamforming and Multiple-input multiple-output communications (MIMO). Beam forming focuses the transmitted power of an antenna in a beam towards the user's direction. MIMO uses multiple antennas at the sending and receiving side. Further releases of the standard have introduced dual carrier operation, i.e. the simultaneous use of two 5 MHz carriers. The technology also delivers significant battery life improvements and dramatically quicker wake-from-idle time, delivering a true always-on connection. HSPA+ is an evolution of HSPA that upgrades the existing 3G network and provides a method for telecom operators to migrate towards 4G speeds that are more comparable to the initially available speeds of newer LTE networks without deploying a new radio interface. HSPA+ should not be confused with LTE though, which uses an air interface based on Orthogonal frequency-division multiple access modulation and multiple access.

Advanced HSPA+ is a further evolution of HSPA+ and provides data rates up to 84.4 and 168 Megabits per second (Mbit/s) to the mobile device (downlink) and 22 Mbit/s from the mobile device (uplink) under ideal signal conditions. Technically these are achieved through the use of a multiple-antenna technique known as MIMO (for "multiple-input and multiple-output") and higher order modulation (64QAM) or combining multiple cells into one with a technique known as Dual-Cell HSDPA.

Downlink

Evolved HSDPA (HSPA+)

An Evolved HSDPA network can theoretically support up to 28 Mbit/s and 42 Mbit/s with a single 5 MHz carrier for Rel7 (MIMO with 16QAM) and Rel8 (64-QAM + MIMO), in good channel conditions with low correlation between transmit antennas. Although real speeds are far lower. Besides the throughput gain from doubling the number of cells to be used, some diversity and joint scheduling gains can also be achieved. The QoS (Quality of Service) can be particularly improved for end users in poor radio reception where they cannot benefit from the other WCDMA capacity improvements (MIMO and higher order modulations) due to poor radio signal quality. In 3GPP a study item was completed in June 2008. The outcome can be found in technical report 25.825. An alternative method to double the data rates is to double the bandwidth to 10 MHz (i.e. 2×5 MHz) by using DC-HSDPA.

Dual-carrier HSDPA (DC-HSDPA)

Dual-Carrier HSDPA, also known as Dual-Cell HSDPA, is part of 3GPP Release 8 specification. It is the natural evolution of HSPA by means of carrier aggregation in the downlink. UMTS licenses are often issued as 5, 10, or 20 MHz paired spectrum allocations. The basic idea of the multicarrier feature is to achieve better resource utilization and spectrum efficiency by means of joint resource allocation and load balancing across the downlink carriers.

New HSDPA User Equipment categories 21-24 have been introduced that support DC-HSDPA. DC-HSDPA can support up to 42.2 Mbit/s, but unlike HSPA, it does not need to rely on MIMO transmission.

The support of MIMO in combination with DC-HSDPA will allow operators deploying Release 7 MIMO to benefit from the DC-HSDPA functionality as defined in Release 8. While in Release 8 DC-HSDPA can only operate on adjacent carriers, Release 9 also allows that the paired cells can operate on two different frequency bands. Later releases allow the use of up to four carriers simultaneously.

From Release 9 onwards it will be possible to use DC-HSDPA in combination with MIMO being used on both carriers. The support of MIMO in combination with DC-HS-DPA will allow operators even more capacity improvements within their network. This will allow theoretical speed of up to 84.4 Mbit/s.

User Equipment (UE) Categories

The following table is derived from table 5.1a of the release 11 of 3GPP TS 25.306 and shows maximum data rates of different device classes and by what combination of features they are achieved. The per-cell per-stream data rate is limited by the *Maximum number of bits of an HS-DSCH transport block received within an HS-DSCH TTI* and the *Minimum inter-TTI interval*. The TTI is 2 ms. So for example Cat 10 can decode 27952 bits/2 ms = 13.976 MBit/s (and not 14.4 MBit/s as often claimed incorrectly). Categories 1-4 and 11 have inter-TTI intervals of 2 or 3, which reduces the maximum data rate by that factor. Dual-Cell and MIMO 2x2 each multiply the maximum data rate by 2, because multiple independent transport blocks are transmitted over different carriers or spatial streams, respectively. The data rates given in the table are rounded to one decimal point.

Uplink

Dual-carrier HSUPA (DC-HSUPA)

Dual-Carrier HSUPA, also known as *Dual-Cell HSUPA*, is a wireless broadband standard based on HSPA that is defined in 3GPP UMTS release 9. Dual Cell (DC-)HSUPA is the natural evolution of HSPA by means of carrier aggregation in the uplink. UMTS licenses are often issued as 10 or 15 MHz paired spectrum allocations. The basic idea of the multi-carrier feature is to achieve better resource utilization and spectrum efficiency by means of joint resource allocation and load balancing across the uplink carriers.

Similar enhancements as introduced with Dual-Cell HSDPA in the downlink for 3GPP Release 8 were standardized for the uplink in 3GPP Release 9, called Dual-Cell HSUPA. The standardisation of Release 9 was completed in December 2009.

User Equipment (UE) Categories

The following table shows uplink speeds for the different categories of Evolved HSUPA.

Multi-Carrier HSPA (MC-HSPA)

The aggregation of more than two carriers has been studied and 3GPP Release 11 is scheduled to include 4-carrier HSPA. The standard is scheduled to be finalised in Q3 2012 and first chipsets supporting MC-HSPA in late 2013. Release 11 specifies 8-carrier HSPA allowed in non-contiguous bands with 4 × 4 MIMO offering peak transfer rates up to 672 Mbit/s.

The 168 Mbit/s and 22 Mbit/s represent theoretical peak speeds. The actual speed for a user will be lower. In general, HSPA+ offers higher bitrates only in very good radio conditions (very close to the cell tower) or if the terminal and network both support either MIMO or Dual-Cell HSDPA, which effectively use two parallel transmit channels with different technical implementations.

The higher 168 Mbit/s speeds are achieved by using multiple carriers with Dual-Cell HSDPA and 4-way MIMO together simultaneously.

All-IP Architecture

A flattened all-IP architecture is an option for the network within HSPA+. In this architecture, the base stations connect to the network via IP (often Ethernet providing the transmission), bypassing legacy elements for the user's data connections. This makes the network faster and cheaper to deploy and operate. The legacy architecture is still permitted with the Evolved HSPA and is likely to exist for several years after adoption of the other aspects of HSPA+ (higher order modulation, multiple streams, etc.).

This 'flat architecture' connects the 'user plane' directly from the base station to the GGSN external gateway, using any available link technology supporting TCP/IP. The definition can be found in 3GPP TR25.999. The user's data flow bypasses the Radio Network Controller (RNC) and the SGSN of the previous 3GPP UMTS architecture versions, thus simplifying the architecture, reducing costs and delays. This is nearly identical to the 3GPP Long Term Evolution (LTE) flat architecture as defined in the 3GPP standard Rel-8. The changes allow cost effective modern link layer technologies such as xDSL or Ethernet, and these technologies are no longer tied to the more expensive and rigid requirements of the older standard of SONET/SDH and E1/T1 infrastructure.

There are no changes to the 'control plane'.

Nokia Siemens Networks Internet HSPA or I-HSPA is the first commercial solution implementing the Evolved HSPA flattened all-IP architecture.

Last mile

The last mile or last kilometer is a colloquial phrase widely used in the telecommunications, cable television and internet industries to refer to the final leg of the telecommunications networks that deliver telecommunication services to retail end-users (customers). More specifically, the last mile refers to the portion of the telecommunications network chain that physically reaches the end-user's premises. Examples are the copper wire subscriber lines connecting landline telephones to the local telephone exchange; coaxial cable service drops carrying cable television signals from utility poles to subscribers' homes, and cell towers linking local cell phones to the cellular network.

The word "mile" is used metaphorically; the length of the last mile link may be more or less than a mile. Because the last mile of a network to the user is conversely the first mile from the user's premises to the outside world when the user is sending data (sending an email, for example), the term first mile is also alternately used.

The last mile is typically the speed bottleneck in communication networks; its bandwidth effectively limits the bandwidth of data that can be delivered to the customer. This is because retail telecommunication networks have the topology of "trees", with relatively few high capacity "trunk" communication channels branching out to feed many final mile "leaves". The final mile links, being the most numerous and thus most expensive part of the system, as well as having to interface with a wide variety of user equipment, are the most difficult to upgrade to new technology. For example, telephone trunklines that carry phone calls between switching centers are made of modern optical fiber, but the last mile is typically twisted pair wires, a technology which has essentially remained unchanged for over a century since the original laying of copper phone cables.

To resolve, or at least mitigate, the problems involved with attempting to provide enhanced services over the last mile, some firms have been mixing networks for decades. One example is fixed wireless access, where a wireless network is used instead of wires to connect a stationary terminal to the wireline network. Various solutions are being developed which are seen as an alternative to the last mile of standard incumbent local exchange carriers. These include WiMAX and broadband over power lines.

In recent years, usage of the term "last mile" has expanded outside the communications industries, to include other distribution networks that deliver goods to customers, such as the pipes that deliver water and natural gas to customer premises, and the final legs of mail and package delivery services.

Business Telephone Service

In many countries the last mile link which connects landline business telephone customers to the local telephone exchange is often an ISDN30 connection delivered through either a copper or fibreoptic cable. This ISDN30 can carry 30 simultaneous telephone calls and many direct dial telephone numbers.

When leaving the telephone exchange, the ISDN30 cable can be buried in the ground, usually in ducting, at very little depth. This makes any business telephone lines vulnerable to being dug up during streetworks, liable to flooding during heavy storms, and effectively subject to general wear and tear due to all manner of natural hazards. Loss, therefore, of the 'last mile' link, means the non-delivery of calls, and other data, to the business affected.

Any business with ISDN30 type connectivity must anticipate such failure in its business continuity planning. There are many options, as documented in customer proprietary network information:

- "Dual parenting" is where the telephone carrier provides the same numbers from two different telephone exchanges. If the cable is damaged from one telephone exchange to the customer premises most of the calls can be delivered from the surviving route to the customer.

- "Diverse routing" is where the carrier can provide more than one route to supply ISDN30 connectivity from the exchange, or exchanges, (as in dual parenting), but they may share underground ducting and cabinets.

- "Separacy" is where the carrier can provide more than one route to bring ISDN30 connectivity from the exchange, or exchanges, (as in dual parenting), but they may not share underground ducting and cabinets, and therefore should be absolutely separate from the telephone exchange to the customer premises.

- "Exchange-based solutions" is where a specialist company working in association with the carriers offers as an enhancement the ability to divert ISDN30 connectivity upon failure to any other number or group of numbers. Carrier diversions are usually limited to all of the ISDN30 direct dial telephone numbers being delivered to one single number.

- "Non-exchange-based diversion services" is where a specialist company working in association with the carrier offers an enhancement to the ability to divert ISDN30 connectivity in case of failure to any other number or group of numbers. Carrier diversions are usually limited to all of the ISDN30 direct dial telephone numbers being delivered to one single number. In the UK Teamphone offers this service in association with British Telecom. By not being in the exchanges, the Teamphone version offers an 'all or nothing' diversion service if required, but does not offer voice recording of calls.

- "Ported number services" is where customers numbers can be ported to a specialist company which points the numbers to the ISDN30 direct dial telephone numbers during business as usual, and delivers them to alternative numbers during a business continuity need. These are generally carrier-independent and there are a number of companies offering such solutions in the UK and AirNorth Communications in the United States.

- "Hosted numbers" is where the carriers or specialist companies can host the customer's numbers within their own or the carrier's networks and deliver calls over an IP network to the customer's sites. When a diversion service is required, the calls can be routed to alternative numbers.

- "Inbound numbers", or "08 type services", is where the carriers or specialist companies can offer 08/05/03 prefixed numbers to the ISDN30 direct dial telephone numbers and can point them to alternative numbers in the event of a diversion requirement. Both carriers and specialist companies offer this type of service in the UK.

Existing Delivery System Problems

The increasing worldwide demand for rapid, low-latency and high-volume communication of information to homes and businesses has made economical information distribution and delivery increasingly important. As demand has escalated, particularly fueled by the widespread adoption of the Internet, the need for economical high-speed access by end-users located at millions of locations has ballooned as well.

As requirements have changed, the existing systems and networks that were initially pressed into service for this purpose have proven to be inadequate. To date, although a number of approaches have been tried, no single clear solution to the 'last mile problem' has emerged.

As expressed by Shannon's equation for channel information capacity, the omnipresence of noise in information systems sets a minimum signal-to-noise ratio (shortened as S/N) requirement in a channel, even when adequate spectral bandwidth is available. Since the integral of the rate of information transfer with respect to time is information quantity, this requirement leads to a corresponding minimum energy per bit. The problem of sending any given amount of information across a channel can therefore be viewed in terms of sending sufficient Information-Carrying Energy (ICE). For this reason the concept of an ICE 'pipe' or 'conduit' is relevant and useful for examining existing systems.

The distribution of information to a great number of widely separated end-users can be compared to the distribution of many other resources. Some familiar analogies are:

- Blood distribution to a large number of cells over a system of veins, arteries and capillaries

- Water distribution by a drip irrigation system to individual plants, including rivers, aqueducts, water mains, etc.

- Nourishment to a plant's leaves through roots, trunk and branches.

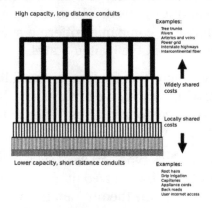

All of these have in common conduits that carry a relatively small amount of a resource a short distance to a very large number of physically separated endpoints. Also common are conduits supporting more voluminous flow, which combine and carry many individual portions over much greater distances. The shorter, lower-volume conduits, which individually serve only one or a small fraction of the endpoints, may have far greater combined length than the larger capacity ones. These common attributes are shown to the right.

Costs and Efficiency

The high-capacity conduits in these systems tend to also have in common the ability to efficiently transfer the resource over a long distance. Only a small fraction of the resource being transferred is wasted, lost, or misdirected. The same cannot necessarily be said of lower-capacity conduits.

One reason has to do with the efficiency of scale. Conduits that are located closer to the endpoint, or end-user, do not individually have as many users supporting them. Even though they are smaller, each has the overhead of an "installation" obtaining and maintaining a suitable path over which the resource can flow. The funding and resources supporting these smaller conduits tend to come from the immediate locale.

This can have the advantage of a "small-government model". That is, the management and resources for these conduits is provided by local entities and therefore can be optimized to achieve the best solutions in the immediate environment and also to make best use of local resources. However, the lower operating efficiencies and relatively greater installation expenses, compared with the transfer capacities, can cause these smaller conduits, as a whole, to be the most expensive and difficult part of the complete distribution system.

These characteristics have been displayed in the birth, growth, and funding of the Internet. The earliest inter-computer communication tended to be accomplished with direct wireline connections between individual computers. These grew into clusters of small local area networks (LAN). The TCP/IP suite of protocols was born out of the need to connect several of these LANs together, particularly as related to common projects among the United States Department of Defense, industry and some academic institutions.

ARPANET came into being to further these interests. In addition to providing a way for multiple computers and users to share a common inter-LAN connection, the TCP/IP protocols provided a standardized way for dissimilar computers and operating systems to exchange information over this inter-network. The funding and support for the connections among LANs could be spread over one or even several LANs.

As each new LAN, or subnet, was added, the new subnet's constituents enjoyed access to the greater network. At the same time the new subnet enabled access to any network or networks with which it was already networked. Thus the growth became a mutually inclusive or "win-win" event.

Economies of Scale

In general, economy of scale makes an increase in capacity of a conduit less expensive as the capacity is increased. There is an overhead associated with the creation of any conduit. This overhead is not repeated as capacity is increased within the potential of the technology being utilized.

As the Internet has grown in size, by some estimates doubling in the number of users every eighteen months, economy of scale has resulted in increasingly large information conduits providing the longest distance and highest capacity backbone connections. In recent years, the capacity of fiber-optic communication, aided by a supporting industry, has resulted in an expansion of raw capacity, so much so that in the United States a large amount of installed fiber infrastructure is not being used because it is currently excess capacity "dark fiber".

This excess backbone capacity exists in spite of the trend of increasing per-user data rates and overall quantity of data. Initially, only the inter-LAN connections were high speed. End-users used existing telephone lines and modems, which were capable of data rates of only a few hundred bit/s. Now almost all end users enjoy access at 100 or more times those early rates.

Economical Information Transfer

Before considering the characteristics of existing last-mile information delivery mechanisms, it is important to further examine what makes information conduits effective. As the Shannon-Hartley theorem shows, it is the combination of bandwidth and signal-to-noise ratio which determines the maximum information rate of a channel. The product of the average information rate and time yields total information transfer. In the presence of noise, this corresponds to some amount of transferred information-carrying energy (ICE). Therefore, the economics of information transfer may be viewed in terms of the economics of the transfer of ICE.

Effective last-mile conduits must:

1. Deliver signal power, S — (must have adequate signal power capacity).

2. Experience low loss (low occurrence of conversion to unusable energy forms).

3. Support wide transmission bandwidth.

4. Deliver high signal-to-noise ratio (SNR) — low unwanted-signal (Noise) power, N.

5. Provide nomadic connectivity.

In addition to these factors, a good solution to the last-mile problem must provide each user:

1. High availability and reliability.

2. Low latency; latency must be small compared with required interaction times.

3. High per-user capacity.

    1. A conduit which is shared among multiple end-users must provide a correspondingly higher capacity in order to properly support each individual user. This must be true for information transfer in each direction.

    2. Affordability; suitable capacity must be financially viable.

Existing Last Mile Delivery Systems

Wired Systems (Including Optical Fiber)

Wired systems provide guided conduits for Information-Carrying Energy (ICE). They all have some degree of shielding, which limits their susceptibility to external noise sources. These transmission lines have losses which are proportional to length. Without the addition of periodic amplification, there is some maximum length beyond which all of these systems fail to deliver an adequate S/N ratio to support information flow. Dielectric optical fiber systems support heavier flow at higher cost.

Local Area Networks (LAN)

Traditional wired local area networking systems require copper coaxial cable or a twisted pair to be run between or among two or more of the nodes in the network. Common systems operate at 100 Mbit/s, and newer ones also support 1000 Mbit/s or more. While length may be limited by collision detection and avoidance requirements, signal loss and reflections over these lines also define a maximum distance. The decrease in information capacity made available to an individual user is roughly proportional to the number of users sharing a LAN.

Telephone

In the late 20th century, improvements in the use of existing copper telephone lines increased their capabilities if maximum line length is controlled. With support for higher transmission bandwidth and improved modulation, these digital subscriber line schemes have increased capability 20-50 times as compared to the previous voiceband systems. These methods are not based on altering the fundamental physical properties and limitations of the medium, which, apart from the introduction of twisted pairs, are no different today than when the first telephone exchange was opened in 1877 by the Bell Telephone Company.

The history and long life of copper-based communications infrastructure is both a testament to the ability to derive new value from simple concepts through technological innovation – and a warning that copper communications infrastructure is beginning to offer diminishing returns for continued investment.

CATV

Community antenna television systems, also known as cable television, have been expanded to provide bidirectional communication over existing physical cables. However, they are by nature shared systems and the spectrum available for reverse information flow and achievable S/N are limited. As was done for initial unidirectional TV communication, cable loss is mitigated through the use of periodic amplifiers within the system. These factors set an upper limit on per-user information capacity, particularly when many users share a common section of cable or access network.

Optical Fiber

Fiber offers high information capacity and after the turn of the 21st century became the deployed medium of choice ("Fiber to the x") given its scalability in the face of the increasing bandwidth requirements of modern applications.

In 2004, according to Richard Lynch, Executive Vice President and Chief Technology Officer of the telecom giant Verizon, the company saw the world moving toward vastly higher bandwidth applications as consumers loved everything broadband had to offer and eagerly devoured as much as they could get, including two-way, user-generated content. Copper and coaxial networks would not – in fact, could not – satisfy these demands, which precipitated Verizon's aggressive move into fiber-to-the-home via FiOS.

Fiber is a future-proof technology that meets the needs of today's users, but unlike other copper-based and wireless last-mile mediums, also has the capacity for years to come, by upgrading the end-point optics and electronics without changing the fiber infrastructure. The fiber itself is installed on existing pole or conduit infrastructure and most of the cost is in labor, providing good regional economic stimulus in the deployment phase and providing a critical foundation for future regional commerce.

Fixed copper lines have been subject to theft due to the value of copper, but optical fibers make unattractive targets. Optical fibers cannot be converted into anything else, whereas copper can be recycled without loss.

Wireless Delivery Systems

Mobile CDN coined the term the 'mobile mile' to categorize the last mile connection when a wireless systems is used to reach the customer. In contrast to wired delivery systems, wireless systems use unguided waves to transmit ICE. They all tend to be unshielded and have a greater degree of susceptibility to unwanted signal and noise sources.

Because these waves are not guided but diverge, in free space these systems are attenuated following an inverse-square law, inversely proportional to distance squared. Losses thus increase more slowly with increasing length than for wired systems, whose loss increases exponentially. In a free space environment, beyond a given length, the losses in a wireless system are lower than those in a wired system.

In practice, the presence of atmosphere, and especially obstructions caused by terrain, buildings and foliage can greatly increase the loss above the free space value. Reflection, refraction and diffraction of waves can also alter their transmission characteristics and require specialized systems to accommodate the accompanying distortions.

Wireless systems have an advantage over wired systems in last mile applications in not requiring lines to be installed. However, they also have a disadvantage in that their unguided nature makes them more susceptible to unwanted noise and signals. Spectral reuse can therefore be limited.

Lightwaves and Free-space Optics

Visible and infrared light waves are much shorter than radio frequency waves. Their use to transmit data is referred to as free-space optical communication. Being short, light waves can be focused or collimated with a small lens/antenna, and to a much higher degree than radio waves. Thus, a receiving device can recover a greater portion of the transmitted signal.

Also, because of the high frequency, a high data transfer rate may be available. However, in practical last mile environments, obstructions and de-steering of these beams, and absorption by elements of the atmosphere including fog and rain, particularly over longer paths, can greatly restrict their use for last-mile wireless communications. Longer (redder) waves suffer less obstruction but may carry lower data rates.

Radio Waves

Radio frequencies (RF), from low frequencies through the microwave region, have wavelengths much longer than visible light. Although this means that it is not possible to focus the beams nearly as tightly as for light, it also means that the aperture or "capture area" of even the simplest, omnidirectional antenna is significantly larger than that of a lens in any feasible optical system. This characteristic results in greatly increased attenuation or "path loss" for systems that are not highly directional.

Actually, the term path loss is something of a misnomer because no energy is lost on a free-space path. Rather, it is merely not received by the receiving antenna. The apparent reduction in transmission, as frequency is increased, is an artifact of the change in the aperture of a given type of antenna.

Relative to the last-mile problem, these longer wavelengths have an advantage over light waves when omnidirectional or sectored transmissions are considered. The larger aperture of radio antennas results in much greater signal levels for a given path length and therefore higher information capacity. On the other hand, the lower carrier frequencies are not able to support the high information bandwidths, which are required by Shannon's equation when the practical limits of S/N have been reached.

For the above reasons, wireless radio systems are optimal for lower-information-capacity broadcast communications delivered over longer paths. For high-information capacity, highly-directive point-to-point over short ranges, wireless light-wave systems are the most useful.

One-way (broadcast) Radio and Television Communications

Historically, most high-information-capacity broadcast has used lower frequencies, generally no higher than the UHF television region, with television itself being a prime example. Terrestrial television has generally been limited to the region above 50 MHz where sufficient information bandwidth is available, and below 1,000 MHz, due to problems associated with increased path loss, as mentioned above.

Two-way Wireless Communications

Two-way communication systems have primarily been limited to lower-information-capacity applications, such as audio, facsimile, or radioteletype. For the most part, higher-capacity systems, such as two-way video communications or terrestrial microwave telephone and data trunks, have been limited and confined to UHF or microwave and to point-point paths.

Higher capacity systems such as third-generation cellular telephone systems require a large infrastructure of more closely spaced cell sites in order to maintain communications within typical environments, where path losses are much greater than in free space and which also require omnidirectional access by the users.

Satellite Communications

For information delivery to end users, satellite systems, by nature, have relatively long path lengths, even for low earth-orbiting satellites. They are also very expensive to deploy and therefore each satellite must serve many users. Additionally, the very long paths of geostationary satellites cause information latency that makes many real-time applications unfeasible.

As a solution to the last-mile problem, satellite systems have application and sharing limitations. The ICE which they transmit must be spread over a relatively large geographical area. This causes the received signal to be relatively small, unless very large or directional terrestrial antennas are used. A parallel problem exists when a satellite is receiving.

In that case, the satellite system must have a very great information capacity in order to accommodate a multitude of sharing users and each user must have large antenna, with attendant directivity and pointing requirements, in order to obtain even modest information-rate transfer. These requirements render high-information-capacity, bi-directional information systems uneconomical. This is one reason why the Iridium satellite system was not more successful.

Broadcast Versus Point-to-point

For terrestrial and satellite systems, economical, high-capacity, last-mile communications requires point-to-point transmission systems. Except for extremely small geographic areas, broadcast systems are only able to deliver high S/N ratios at low frequencies where there is not sufficient spectrum to support the large information capacity needed by a large number of users. Although complete "flooding" of a region can be accomplished, such systems have the fundamental characteristic that most of the radiated ICE never reaches a user and is wasted.

As information requirements increase, broadcast wireless mesh systems (also sometimes referred to as microcells or nano-cells) which are small enough to provide adequate information distribution to and from a relatively small number of local users require a prohibitively large number of broadcast locations or points of presence along with a large amount of excess capacity to make up for the wasted energy.

Intermediate System

Recently a new type of information transport midway between wired and wireless systems has been discovered. Called E-Line, it uses a single central conductor but no outer conductor or shield. The energy is transported in a plane wave which, unlike radio does not diverge, whereas like radio it has no outer guiding structure.

This system exhibits a combination of the attributes of wired and wireless systems and can support high information capacity utilizing existing power lines over a broad range of frequencies from RF through microwave.

Line Aggregation

Aggregation is a method of bonding multiple lines to achieve a faster, more reliable connection. Some companies believe that ADSL aggregation (or "bonding") is the solution to the UK's last mile problem.

Wireless Local Loop

Wireless local loop (WLL), is the use of a wireless communications link as the "last mile

/ first mile" connection for delivering plain old telephone service (POTS) or Internet access (marketed under the term "broadband") to telecommunications customers. Various types of WLL systems and technologies exist.

Other terms for this type of access include Broadband Wireless Access (BWA), Radio In The Loop (RITL), Fixed-Radio Access (FRA), Fixed Wireless Access (FWA) and Metro Wireless (MW).

Definition of Fixed Wireless Service

Fixed Wireless Terminal (FWT) units differ from conventional mobile terminal units operating within cellular networks – such as GSM – in that a fixed wireless terminal or desk phone will be limited to an almost permanent location with almost no roaming abilities.

WLL and FWT are generic terms for radio based telecommunications technologies and the respective devices which can be implemented using a number of different wireless and radio technologies.

Wireless local loop services are segmented into a number of broad market and deployment groups. Services are split between licensed – commonly used by carriers and Telcos – and unlicensed services more commonly deployed by home users and Wireless ISPs (WISPs).

Licensed Points-to-point Microwave Services

Licensed microwave services have been used since the 1960s to transmit very large amounts of data. The AT&T Long Lines coast to coast backbone in the USA was largely carried over a chain of microwave towers. These systems have been largely using 3700–4200 MHz and 5000–6200 MHz. The 5 GHz band was even known as the "common carrier" band. This service typically was prohibitively expensive to be used for local loops, and was used for backbone networks. In the 1980s and 1990s it flourished under the growth of cell towers. This growth spurred research in this area, and as the cost continues to decline, it is being used as an alternative to T-1, T-3, and fiber connectivity.

Licensed Point-to-multipoint Microwave Services

Multipoint microwave licenses are generally more expensive than point to point licenses. A single point to point system could be installed and licensed for 50,000 to 200,000 USD. A multipoint license would start in the millions of dollars. Multichannel Multipoint Distribution Service (MMDS) and Local Multipoint Distribution Service (LMDS) were the first true multi point services for wireless local loop. While Europe and the rest of the world developed the 3500 MHz band for affordable broadband fixed wireless, the U.S. provided LMDS and MMDS, and most implementations in the United States were conducted at 2500 MHz. The largest was Sprint Broadband's deployment of Hybrid

Networks equipment. Sprint was plagued with difficulties operating the network profitably, and service was often spotty, due to inadequate radio link quality.

Unlicensed Multi Point Wireless Service

Most of the growth in long range radio communications since 2002 has been in the license free bands (mostly 900 MHz, 2.4 GHz and 5.8 GHz). Global Pacific Internet and Innetix started wireless service in California in 1995 using Breezecom (Alvarion) frequency hopping radio which later became the standard 802.11.

A few years later NextWeb Networks of Fremont began deploying reliable license free service. For Nextweb they originally deployed 802.11b equipment and later switched to Axxcelera which uses propriety protocol.

1995–2004: License-free Equipment

Most of the early vendors of license-free fixed wireless equipment such as Adaptive Broadband (Axxcelera), Trango Broadband, Motorola (Orthogon), Proxim Networks, Redline Communications and BreezeCom (Alvarion) used proprietary protocols and hardware, creating pressure on the industry to adopt a standard for unlicensed fixed wireless. These Mac Layers typically used a 15–20 MHz channel using Direct Sequence Spread Spectrum and BPSK, CCK and QPSK for modulation.

These devices all describe the customer premises wireless system as the Subscriber Unit (SU), and the operator transmitter delivering the last mile local loop services as the Access Point (AP). 802.11 uses the terms AP and STA (Station).

2002–2005: Wi-Fi Local Loop

Originally designed for short range mobile internet and local area network access, IEEE 802.11 has emerged as the de facto standard for unlicensed Wireless Local Loop. More 802.11 equipment is deployed for long range data service than any other technology. These systems have provided varying results, as the operators were often small and poorly trained in radio communications, additionally 802.11 was not intended to be used at long ranges and suffered from a number of problems, such as the hidden node problem. Many companies such as KarlNet began modifying the 802.11 MAC to attempt to deliver higher performance at long ranges.

2005–present: Maturation of the Wireless ISP Market

In nearly every metropolitan area worldwide, operators and hobbyists deployed more and more unlicensed broadband point to multipoint systems. Providers that had rave reviews when they started faced the prospect of seeing their networks degrade in performance, as more and more devices were deployed using the license free U-NII (5.3/5.4 GHz) and ISM (2.4 and 5.8 GHz) bands and competitors sprung up around them.

The Growing Interference Problem

Interference caused the majority of unlicensed wireless services to have much higher error rates and interruptions than equivalent wired or licensed wireless networks, such as the copper telephone network, and the coaxial cable network. This caused growth to slow, customers to cancel, and many operators to rethink their business model.

There were several responses to these problems.

2003: Voluntary Frequency Coordination (USA)

Next-Web, Etheric Networks, Gate Speed and a handful of other companies founded the first voluntary spectrum coordination body – working entirely independently of government regulators. This organization was founded in March 2003 as BANC, "Bay Area Network Coordination". By maintaining frequencies used in an inter-operator database, disruptions between coordinating parties were minimized, as well as the cost of identifying new or changing transmission sources, by using the frequency database to determine what bands were in use. Because the parties in BANC comprised the majority of operators in the Bay Area, they used peer pressure to imply that operators who did not play nice would be collectively punished by the group, through interfering with the non cooperative, while striving not to interfere with the cooperative. BANC was then deployed in Los Angeles. Companies such as Deutsche Telekom joined. It looked like the idea had promise.

2005: Operators Flee Unlicensed for Licensed

The better capitalized operators began reducing their focus on unlicensed and instead focused on licensed systems, as the constant fluctuations in signal quality caused them to have very high maintenance costs. NextWeb, acquired by Covad for a very small premium over the capital invested in it, is one operator who focused on licensed service, as did WiLine Networks. This led to fewer of the more responsible and significant operators actually using the BANC system. Without its founders active involvement, the system languished.

2005 to present: Adaptive Network Technology

Operators began to apply the principles of self-healing networks. Etheric Networks followed this path. Etheric Networks focused on improving performance by developing dynamic interference and fault detection and reconfiguration, as well as optimizing quality based routing software, such as MANET and using multiple paths to deliver service to customers. This approach is generally called "mesh networking" which relies on ad hoc networking protocols, however mesh and ad hoc networking protocols have yet to deliver high speed low latency business class end to end reliable local loop service, as the paths can sometimes traverse exponentially more radio links than a traditional star (AP->SU) topology.

Adaptive network management actively monitors the local loop quality and behaviour, using automation to reconfigure the network and its traffic flows, to avoid interference and other failures.

Mobile Technologies

These are available in Code Division Multiple Access(CDMA), Digital Enhanced Cord-less Telecommunications – DECT (TDMA/DCA) (ETSI 6 EN 300 765-1 V1.3.1 (2001–04) -"Digital Enhanced Cordless Telecommunications (DECT); Radio in the Lo-cal Loop (RLL) Access Profile (RAP); Part 1: Basic telephony services"), Global System for Mobile Communications(GSM), IS136 Time Division Multiple Access (TDMA) as well as analog access technologies such as Advanced Mobile Phone System(AMPS), for which there will be independent standards defining every aspect of modulation, protocols, error handling, etc.

Deployment

The Wireless Local Loop market is currently an extremely high growth market, offering internet service providers immediate access to customer markets without having to either lay cable through a metropolitan area MTA, or work through the ILECs, reselling the telephone, cable or satellite networks, owned by companies that prefer to largely sell direct.

This trend revived the prospects for local and regional ISPs, as those willing to deploy fixed wireless networks were not at the mercy of the large telecommunication monop-olies. They were at the mercy of unregulated re-use of unlicensed frequencies upon which they communicate.

Due to the enormous quantity of 802.11 "Wi-Fi" equipment and software, coupled with the fact that spectrum licenses are not required in the ISM and U-NII bands, the indus-try has moved well ahead of the regulators and the standards bodies.

In 2008, Sprint and ClearWire were preparing to roll out massive WiMAX networks in the United States, but those talks may be stalled pending new investment.

Customer-premises Equipment

Customer-premises equipment or customer-provided equipment (CPE) is any termi-nal and associated equipment located at a subscriber's premises and connected with a carrier's telecommunication channel at the demarcation point ("demarc"). The demarc is a point established in a building or complex to separate customer equipment from the equipment located in either the distribution infrastructure or central office of the communications service provider.

CPE generally refers to devices such as telephones, routers, switches, residential gateways (RG), set-top boxes, fixed mobile convergence products, home networking adapters and Internet access gateways that enable consumers to access communications service providers' services and distribute them around their house via a local area network (LAN).

A CPE can be an active equipment, as the ones mentioned above or a passive equipment such as analogue-telephone-adapters or xDSL-splitters.

Included are key telephone systems and most private branch exchanges. Excluded from CPE are overvoltage protection equipment and pay telephones. Other types of materials that are necessary for the delivery of the telecommunication service, but are not defined as equipment, such as manuals and cable packages, and cable adapters are instead referred to as CPE-peripherals.

CPE can refer to devices purchased by the subscriber, or to those provided by the operator or service provider.

History

The two phrases, "customer-*premises* equipment" and "customer-*provided* equipment", reflect the history of this equipment.

Under the Bell System monopoly in the United States (post Communications Act of 1934), the Bell System owned the phones, and one could not attach one's own devices to the network, or even attach anything to the phones. Thus phones were property of the Bell System, located on customers' premises - hence, customer-*premises* equipment. In the U.S. Federal Communications Commission (FCC) proceeding the Second Computer Inquiry, the FCC ruled that telecommunications carriers could no longer bundle CPE with telecommunications service, uncoupling the market power of the telecommunications service monopoly from the CPE market, and creating a competitive CPE market.

With the gradual breakup of the Bell monopoly, starting with Hush-A-Phone v. United States [1956], which allowed some non-Bell owned equipment to be connected to the network (a process called interconnection), equipment on customers' premises became increasingly owned by customers, not the telco. Indeed, one eventually became able to purchase one's own phone - hence, customer-*provided* equipment.

In the Pay TV industry many operators and service providers offer subscribers a set-top box with which to receive video services, in return for a monthly fee. As offerings have evolved to include multiple services [voice and data] operators have increasingly given consumers the opportunity to rent or buy additional devices like access modems, internet gateways and video extenders that enable them to access multiple services, and distribute them to a range of Consumer Electronics devices around the home.

Technology Evolution

Hybrid Devices

The growth of multiple-service operators, offering triple or quad-play services, required the development of hybrid CPE to make it easy for subscribers to access voice, video and data services. The development of this technology was led by Pay TV operators looking for a way to deliver video services via both traditional broadcast and broadband IP networks. Spain's Telefonica was the first operator to launch a hybrid broadcast and broadband TV service in 2003 with its Movistar TV DTT/IPTV offering, while Polish satellite operator 'n' was the first to offer its subscribers a Three-way hybrid (or Tri-brid) broadcast and broadband TV service, which launched in 2009

Set-back Box

The term set-back box is used in the digital TV industry to describe a piece of consumer hardware that enables them to access both linear broadcast and internet-based video content, plus a range of interactive services like Electronic Programme Guides (EPG), Pay Per View (PPV) and video on demand (VOD) as well as internet browsing, and view them on a large screen television set. Unlike standard set-top boxes, which sit on top of or below the TV, a set-back box has a smaller form factor to enable it to be mounted to the rear of the display panel flat panel TV, hiding it from view.

Home Gateway

A residential gateway is a home networking device used to connect devices in the home to the Internet or other WAN. It is an umbrella term, used to cover multi-function networking appliances used in homes, which may combine a DSL modem or cable modem, a network switch, a consumer-grade router, and a wireless access point. In the past, such functions were provided by separate devices, but in recent years technological convergence has enabled multiple functions to merged into a single device.

One of the first home gateway devices to be launched was selected by Telecom Italia to enable the operator to offer triple play services in 2002 . Along with a SIP VoIP handset for making voice calls, it enabled subscribers to access voice, video and data services over a 10MB symmetrical ADSL fiber connection.

Virtual Gateway

The virtual gateway concept enables consumers to access video and data services and distribute them around their homes using software rather than hardware. The first virtual gateway was introduced in 2010 by Advanced Digital Broadcast at the IBC exhibition in Amsterdam. The ADB Virtual Gateway uses software that resides within the middleware and is based on open standards, including DLNA home networking and the DTCP-IP standard, to ensure that all content, including paid-for encrypted content like Pay TV services, can only be accessed by secure CE devices.

Broadband

A subscriber unit, or SU is a broadband radio that is installed at a business or residential location to connect to an access point to send/receive high speed data wired or wirelessly. Devices commonly referred to as a subscriber unit include cable modems, access gateways, home networking adapters and mobile phones.

WAN

The terms "customer-premises equipment", "customer-provided equipment", or "CPE" may also refer to any devices that terminate a WAN circuit, such as an ISDN, E-carrier/T-carrier, DSL, or metro Ethernet. This includes any customer-owned hardware at the customer's site: routers, firewalls, network switches, PBXs, VoIP gateways, sometimes CSU/DSU and modems.

Application Areas

- Connected Home
- Pay TV
- Over-the-top video services
- Broadband
- Voice over IP
- Fixed–mobile convergence [FMC]

Other Uses

- Cellular carriers may sometimes internally refer to cellular phones a customer has purchased without a subsidy or from a third party as "customer provided equipment."

- It is also notable that the fully qualified domain name and the PTR record of DSL and cable lines connected to a residence will often contain 'cpe'.

References

- "Universal Mobile Telecommunications System (UMTS); UE Radio Access capabilities" (PDF). ETSI. January 2014. Retrieved March 4, 2014.

- Klas Johansson; Johan Bergman; Dirk Gerstenberger; Mats Blomgren; Anders Wallén (28 January 2009). "Multi-Carrier HSPA Evolution" (PDF). Ericsson.com. Retrieved 2014-06-01.

- "White paper Long Term HSPA Evolution Mobile broadband evolution beyond 3GPP Release 10" (PDF). Nokiaslemensnetworks.com. 14 December 2010. Retrieved 2014-06-01.

- "Cable Sharing in Commercial Building Environments: Reducing Cost, Simplifying Cable Management, and Converging Applications onto Twisted-Pair Media". Siemon.com. Retrieved 2014-04-28.

- "Ericsson Review #1 2009 - Continued HSPA Evolution of mobile broadband" (PDF). Ericsson.com. 27 January 2009. Retrieved 2014-06-01.

- "The Evolution of Copper Cabling Systems from Cat5 to Cat5e to Cat6" (PDF). Panduit. 2004-02-27. Retrieved 2013-05-12.

- "Cat5 Spec, cat6 specs, cat7 spec - Definitions, Comparison, Specifications". TEC Datawire. Retrieved 2013-01-05"."

- "HDBaseT Alliance Shows the Future of Connected Home Entertainment at CES 2013" (PDF). HDBaset.org. January 9, 2013. Retrieved June 5, 2013.

- "GSA confirms 70% jump in 42 Mbit/s DC-HSPA+ network deployments over past 3 months". Gsacom.com. Retrieved 2012-11-27.

Competing Technologies of WiMAX

There are several wireless technologies that are contending with WiMAX some of which are Local Multipoint Distribution Service, Wireless broadband, LTE (telecommunication), CDMA2000, UMTS (telecommunication) etc. WiMAX has distinct advantages over the other competing technologies and this chapter is a comparative study of these advantages.

UMTS (Telecommunication)

The Universal Mobile Telecommunications System (UMTS) is a third generation mobile cellular system for networks based on the GSM standard. Developed and maintained by the 3GPP (3rd Generation Partnership Project), UMTS is a component of the International Telecommunications Union IMT-2000 standard set and compares with the CDMA2000 standard set for networks based on the competing cdmaOne technology. UMTS uses wideband code division multiple access (W-CDMA) radio access technology to offer greater spectral efficiency and bandwidth to mobile network operators.

UMTS specifies a complete network system, which includes the radio access network (UMTS Terrestrial Radio Access Network, or UTRAN), the core network (Mobile Application Part, or MAP) and the authentication of users via SIM (subscriber identity module) cards.

The technology described in UMTS is sometimes also referred to as Freedom of Mobile Multimedia Access (FOMA) or 3GSM.

Unlike EDGE (IMT Single-Carrier, based on GSM) and CDMA2000 (IMT Multi-Carrier), UMTS requires new base stations and new frequency allocations.

Features

UMTS supports maximum theoretical data transfer rates of 42 Mbit/s when Evolved HSPA (HSPA+) is implemented in the network. Users in deployed networks can expect a transfer rate of up to 384 kbit/s for Release '99 (R99) handsets (the original UMTS release), and 7.2 Mbit/s for High-Speed Downlink Packet Access (HSDPA) handsets in the downlink connection. These speeds are significantly faster than the 9.6 kbit/s of a single GSM error-corrected circuit switched data channel, multiple 9.6 kbit/s channels in High-Speed Circuit-Switched Data (HSCSD) and 14.4 kbit/s for CDMAOne channels.

Since 2006, UMTS networks in many countries have been or are in the process of being upgraded with High-Speed Downlink Packet Access (HSDPA), sometimes known as 3.5G. Currently, HSDPA enables downlink transfer speeds of up to 21 Mbit/s. Work is also progressing on improving the uplink transfer speed with the High-Speed Uplink Packet Access (HSUPA). Longer term, the 3GPP Long Term Evolution (LTE) project plans to move UMTS to 4G speeds of 100 Mbit/s down and 50 Mbit/s up, using a next generation air interface technology based upon orthogonal frequency-division multiplexing.

The first national consumer UMTS networks launched in 2002 with a heavy emphasis on telco-provided mobile applications such as mobile TV and video calling. The high data speeds of UMTS are now most often utilised for Internet access: experience in Japan and elsewhere has shown that user demand for video calls is not high, and telco-provided audio/video content has declined in popularity in favour of high-speed access to the World Wide Web—either directly on a handset or connected to a computer via Wi-Fi, Bluetooth or USB.

Air Interfaces

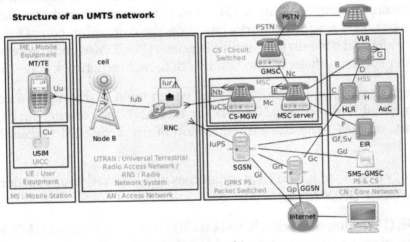

UMTS network architecture

UMTS combines three different terrestrial air interfaces, GSM's Mobile Application Part (MAP) core, and the GSM family of speech codecs.

The air interfaces are called UMTS Terrestrial Radio Access (UTRA). All air interface options are part of ITU's IMT-2000. In the currently most popular variant for cellular mobile telephones, W-CDMA (IMT Direct Spread) is used.

Please note that the terms W-CDMA, TD-CDMA and TD-SCDMA are misleading. While they suggest covering just a channel access method (namely a variant of CDMA), they are actually the common names for the whole air interface standards.

W-CDMA (UTRA-FDD)

3G sign shown in notification bar on an Android powered smartphone.

W-CDMA or WCDMA (Wideband Code Division Multiple Access), along with UMTS-FDD, UTRA-FDD, or IMT-2000 CDMA Direct Spread is an air interface standard found in 3G mobile telecommunications networks. It supports conventional cellular voice, text and MMS services, but can also carry data at high speeds, allowing mobile operators to deliver higher bandwidth applications including streaming and broadband Internet access.

UMTS base station on the roof of a building

W-CDMA uses the DS-CDMA channel access method with a pair of 5 MHz wide channels. In contrast, the competing CDMA2000 system uses one or more available 1.25 MHz channels for each direction of communication. W-CDMA systems are widely criticized for their large spectrum usage, which delayed deployment in countries that acted relatively slowly in allocating new frequencies specifically for 3G services (such as the United States).

The specific frequency bands originally defined by the UMTS standard are 1885–2025 MHz for the mobile-to-base (uplink) and 2110–2200 MHz for the base-to-mobile (downlink). In the US, 1710–1755 MHz and 2110–2155 MHz are used instead, as the 1900 MHz band was already used. While UMTS2100 is the most widely deployed UMTS band, some countries' UMTS operators use the 850 MHz and/or 1900 MHz bands (independently, meaning uplink and downlink are within the same band), notably in the US by AT&T Mobility, New Zealand by Telecom New Zealand on the XT Mobile Network and in Australia by Telstra on the Next G network. Some carriers such as T-Mobile use band numbers to identify the UMTS frequencies. For example, Band I (2100 MHz), Band IV (1700/2100 MHz), and Band V (850 MHz).

UMTS-FDD is an acronym for Universal Mobile Telecommunications System (UMTS) - frequency-division duplexing (FDD) and a 3GPP standardized version of UMTS networks that makes use of frequency-division duplexing for duplexing over an UMTS Terrestrial Radio Access (UTRA) air interface.

W-CDMA is the basis of Japan's NTT DoCoMo's FOMA service and the most-commonly used member of the Universal Mobile Telecommunications System (UMTS) family and sometimes used as a synonym for UMTS. It uses the DS-CDMA channel access method and the FDD duplexing method to achieve higher speeds and support more users compared to most previously used time division multiple access (TDMA) and time division duplex (TDD) schemes.

While not an evolutionary upgrade on the airside, it uses the same core network as the 2G GSM networks deployed worldwide, allowing dual mode mobile operation along with GSM/EDGE; a feature it shares with other members of the UMTS family.

Development

In the late 1990s, W-CDMA was developed by NTT DoCoMo as the air interface for their 3G network FOMA. Later NTT DoCoMo submitted the specification to the International Telecommunication Union (ITU) as a candidate for the international 3G standard known as IMT-2000. The ITU eventually accepted W-CDMA as part of the IMT-2000 family of 3G standards, as an alternative to CDMA2000, EDGE, and the short range DECT system. Later, W-CDMA was selected as an air interface for UMTS.

As NTT DoCoMo did not wait for the finalisation of the 3G Release 99 specification, their network was initially incompatible with UMTS. However, this has been resolved by NTT DoCoMo updating their network.

Code Division Multiple Access communication networks have been developed by a number of companies over the years, but development of cell-phone networks based on CDMA (prior to W-CDMA) was dominated by Qualcomm, the first company to succeed in developing a practical and cost-effective CDMA implementation for consumer cell phones and its early IS-95 air interface standard has evolved into the current CDMA2000 (IS-856/IS-2000) standard. Qualcomm created an experimental wideband CDMA system called CDMA2000 3x which unified the W-CDMA (3GPP) and CDMA2000 (3GPP2) network technologies into a single design for a worldwide standard air interface. Compatibility with CDMA2000 would have beneficially enabled roaming on existing networks beyond Japan, since Qualcomm CDMA2000 networks are widely deployed, especially in the Americas, with coverage in 58 countries as of

2006. However, divergent requirements resulted in the W-CDMA standard being retained and deployed globally. W-CDMA has then become the dominant technology with 457 commercial networks in 178 countries as of April 2012. Several CDMA2000 operators have even converted their networks to W-CDMA for international roaming compatibility and smooth upgrade path to LTE.

Despite incompatibility with existing air-interface standards, late introduction and the high upgrade cost of deploying an all-new transmitter technology, W-CDMA has become the dominant standard.

Rationale for W-CDMA

W-CDMA transmits on a pair of 5 MHz-wide radio channels, while CDMA2000 transmits on one or several pairs of 1.25 MHz radio channels. Though W-CDMA does use a direct sequence CDMA transmission technique like CDMA2000, W-CDMA is not simply a wideband version of CDMA2000. The W-CDMA system is a new design by NTT DoCoMo, and it differs in many aspects from CDMA2000. From an engineering point of view, W-CDMA provides a different balance of trade-offs between cost, capacity, performance, and density; it also promises to achieve a benefit of reduced cost for video phone handsets. W-CDMA may also be better suited for deployment in the very dense cities of Europe and Asia. However, hurdles remain, and cross-licensing of patents between Qualcomm and W-CDMA vendors has not eliminated possible patent issues due to the features of W-CDMA which remain covered by Qualcomm patents.

W-CDMA has been developed into a complete set of specifications, a detailed protocol that defines how a mobile phone communicates with the tower, how signals are modulated, how datagrams are structured, and system interfaces are specified allowing free competition on technology elements.

Deployment

The world's first commercial W-CDMA service, FOMA, was launched by NTT DoCoMo in Japan in 2001.

Elsewhere, W-CDMA deployments are usually marketed under the UMTS brand.

W-CDMA has also been adapted for use in satellite communications on the U.S. Mobile User Objective System using geosynchronous satellites in place of cell towers.

J-Phone Japan (once Vodafone and now SoftBank Mobile) soon followed by launching their own W-CDMA based service, originally branded "Vodafone Global Standard" and claiming UMTS compatibility. The name of the service was changed to "Vodafone 3G" (now "SoftBank 3G") in December 2004.

Beginning in 2003, Hutchison Whampoa gradually launched their upstart UMTS networks.

Most countries have, since the ITU approved of the 3G mobile service, either "auctioned" the radio frequencies to the company willing to pay the most, or conducted a "beauty contest"—asking the various companies to present what they intend to commit to if awarded the licences. This strategy has been criticised for aiming to drain the cash of operators to the brink of bankruptcy in order to honour their bids or proposals. Most of them have a time constraint for the rollout of the service—where a certain "coverage" must be achieved within a given date or the licence will be revoked.

Vodafone launched several UMTS networks in Europe in February 2004. MobileOne of Singapore commercially launched its 3G (W-CDMA) services in February 2005. New Zealand in August 2005 and Australia in October 2005.

AT&T Wireless (now a part of Cingular Wireless) has deployed UMTS in several cities. Though advancements in its network deployment have been delayed due to the merger with Cingular, Cingular began offering HSDPA service in December 2005.

Rogers in Canada March 2007 has launched HSDPA in the Toronto Golden Horseshoe district on W-CDMA at 850/1900 MHz and plan the launch the service commercial in the top 25 cities October, 2007.

TeliaSonera opened W-CDMA service in Finland October 13, 2004 with speeds up to 384 kbit/s. Availability only in main cities. Pricing is approx. €2/MB.

SK Telecom and KTF, two largest mobile phone service providers in South Korea, have each started offering W-CDMA service in December 2003. Due to poor coverage and lack of choice in handhelds, the W-CDMA service has barely made a dent in the Korean market which was dominated by CDMA2000. By October 2006 both companies are covering more than 90 cities while SK Telecom has announced that it will provide nationwide coverage for its WCDMA network in order for it to offer SBSM (Single Band Single Mode) handsets by the first half of 2007. KT Freecel will thus cut funding to its CDMA2000 network development to the minimum.

In Norway, Telenor introduced W-CDMA in major cities by the end of 2004, while their competitor, NetCom, followed suit a few months later. Both operators have 98% national coverage on EDGE, but Telenor has parallel WLAN roaming networks on GSM, where the UMTS service is competing with this. For this reason Telenor is dropping support of their WLAN service in Austria (2006).

Maxis Communications and Celcom, two mobile phone service providers in Malaysia, started offering W-CDMA services in 2005.

In Sweden, Telia introduced W-CDMA March 2004.

UTRA-TDD

UMTS-TDD, an acronym for Universal Mobile Telecommunications System (UMTS) -

time-division duplexing (TDD), is a 3GPP standardized version of UMTS networks that use UTRA-TDD. UTRA-TDD is a UTRA that uses time-division duplexing for duplexing. While a full implementation of UMTS, it is mainly used to provide Internet access in circumstances similar to those where WiMAX might be used. UMTS-TDD is not directly compatible with UMTS-FDD: a device designed to use one standard cannot, unless specifically designed to, work on the other, because of the difference in air interface technologies and frequencies used. It is more formally as IMT-2000 CDMA-TDD or IMT 2000 Time-Division (IMT-TD).

The two UMTS air interfaces (UTRAs) for UMTS-TDD are TD-CDMA and TD-SCDMA. Both air interfaces use a combination of two channel access methods, code division multiple access (CDMA) and time division multiple access (TDMA): the frequency band is divided into time slots (TDMA), which are further divided into channels using CDMA spreading codes. These air interfaces are classified as TDD, because time slots can be allocated to either uplink or downlink traffic.

TD-CDMA (UTRA-TDD 3.84 Mcps High Chip Rate (HCR))

TD-CDMA, an acronym for Time-division-Code division multiple access, is a channel access method based on using spread spectrum multiple access (CDMA) across multiple time slots (TDMA). TD-CDMA is the channel access method for UTRA-TDD HCR, which is an acronym for UMTS Terrestrial Radio Access-Time Division Duplex High Chip Rate.

UMTS-TDD's air interfaces that use the TD-CDMA channel access technique are standardized as UTRA-TDD HCR, which uses increments of 5 MHz of spectrum, each slice divided into 10 ms frames containing fifteen time slots (1500 per second). The time slots (TS) are allocated in fixed percentage for downlink and uplink. TD-CDMA is used to multiplex streams from or to multiple transceivers. Unlike W-CDMA, it does not need separate frequency bands for up- and downstream, allowing deployment in tight frequency bands.

TD-CDMA is a part of IMT-2000, defined as IMT-TD Time-Division (IMT CDMA TDD), and is one of the three UMTS air interfaces (UTRAs), as standardized by the 3GPP in UTRA-TDD HCR. UTRA-TDD HCR is closely related to W-CDMA, and provides the same types of channels where possible. UMTS's HSDPA/HSUPA enhancements are also implemented under TD-CDMA.

TD-SCDMA (UTRA-TDD 1.28 Mcps Low Chip Rate (LCR))

Time Division Synchronous Code Division Multiple Access (TD-SCDMA) or UTRA TDD 1.28 mcps low chip rate (UTRA-TDD LCR) is an air interface found in UMTS mobile telecommunications networks in China as an alternative to W-CDMA.

TD-SCDMA uses the TDMA channel access method combined with an adaptive synchronous CDMA component on 1.6 MHz slices of spectrum, allowing deployment in even tighter frequency bands than TD-CDMA. It is standardized by the 3GPP and also referred to as "UTRA-TDD LCR" However, the main incentive for development of this Chinese-developed standard was avoiding or reducing the license fees that have to be paid to non-Chinese patent owners. Unlike the other air interfaces, TD-SCDMA was not part of UMTS from the beginning but has been added in Release 4 of the specification.

Like TD-CDMA, TD-SCDMA is known as IMT CDMA TDD within IMT-2000.

The term "TD-SCDMA" is misleading. While it suggests covering only a channel access method, it is actually the common name for the whole air interface specification.

TD-SCDMA / UMTS-TDD (LCR) networks are incompatible with W-CDMA / UMTS-FDD and TD-CDMA / UMTS-TDD (HCR) networks.

Objectives

TD-SCDMA was developed in the People's Republic of China by the Chinese Academy of Telecommunications Technology (CATT), Datang Telecom, and Siemens AG in an attempt to avoid dependence on Western technology. This is likely primarily for practical reasons, since other 3G formats require the payment of patent fees to a large number of Western patent holders.

TD-SCDMA proponents also claim it is better suited for densely populated areas. Further, it is supposed to cover all usage scenarios, whereas W-CDMA is optimised for symmetric traffic and macro cells, while TD-CDMA is best used in low mobility scenarios within micro or pico cells.

TD-SCDMA is based on spread spectrum technology which makes it unlikely that it will be able to completely escape the payment of license fees to western patent holders. The launch of a national TD-SCDMA network was initially projected by 2005 but only reached large scale commercial trials with 60,000 users across eight cities in 2008.

On January 7, 2009, China granted a TD-SCDMA 3G licence to China Mobile.

On September 21, 2009, China Mobile officially announced that it had 1,327,000 TD-SCDMA subscribers as of the end of August, 2009.

While TD is primarily a China-only system, it may well be exported to developing countries. It is likely to be replaced with a newer TD-LTE system over the next 5 years.

Technical Highlights

TD-SCDMA uses TDD, in contrast to the FDD scheme used by W-CDMA. By dynamically adjusting the number of timeslots used for downlink and uplink, the system can more easily accommodate asymmetric traffic with different data rate requirements on downlink and uplink than FDD schemes. Since it does not require paired spectrum for downlink and uplink, spectrum allocation flexibility is also increased. Using the same carrier frequency for uplink and downlink also means that the channel condition is the same on both directions, and the base station can deduce the downlink channel information from uplink channel estimates, which is helpful to the application of beamforming techniques.

TD-SCDMA also uses TDMA in addition to the CDMA used in WCDMA. This reduces the number of users in each timeslot, which reduces the implementation complexity of multiuser detection and beamforming schemes, but the non-continuous transmission also reduces coverage (because of the higher peak power needed), mobility (because of lower power control frequency) and complicates radio resource management algorithms.

The "S" in TD-SCDMA stands for "synchronous", which means that uplink signals are synchronized at the base station receiver, achieved by continuous timing adjustments. This reduces the interference between users of the same timeslot using different codes by improving the orthogonality between the codes, therefore increasing system capacity, at the cost of some hardware complexity in achieving uplink synchronization.

History

On January 20, 2006, Ministry of Information Industry of the People's Republic of China formally announced that TD-SCDMA is the country's standard of 3G mobile telecommunication. On February 15, 2006, a timeline for deployment of the network in China was announced, stating pre-commercial trials would take place starting after completion of a number of test networks in select cities. These trials ran from March to October, 2006, but the results were apparently unsatisfactory. In early 2007, the Chinese government instructed the dominant cellular carrier, China Mobile, to build commercial trial networks in eight cities, and the two fixed-line carriers, China Telecom and China Netcom, to build one each in two other cities. Construction of these trial networks was scheduled to finish during the fourth quarter of 2007, but delays meant that construction was not complete until early 2008.

The standard has been adopted by 3GPP since Rel-4, known as "UTRA TDD 1.28Mbps Option".

On March 28, 2008, China Mobile Group announced TD-SCDMA "commercial trials" for 60,000 test users in eight cities from April 1, 2008. Networks using other 3G standards (WCDMA and CDMA2000 EV/DO) had still not been launched in China, as these were delayed until TD-SCDMA was ready for commercial launch.

In January 2009 the Ministry of Industry and Information Technology (MIIT) in China took the unusual step of assigning licences for 3 different third-generation mobile phone standards to three carriers in a long-awaited step that is expected to prompt $41 billion in spending on new equipment. The Chinese-developed standard, TD-SCDMA, was assigned to China Mobile, the world's biggest phone carrier by subscribers. That appeared to be an effort to make sure the new system has the financial and technical backing to succeed. Licences for two existing 3G standards, W-CDMA and CDMA2000 1xEV-DO, were assigned to China Unicom and China Telecom, respectively. Third-generation, or 3G, technology supports Web surfing, wireless video and other services and the start of service is expected to spur new revenue growth.

Frequency Bands & Deployments

The following is a list of mobile telecommunications networks using third-generation TD-SCDMA / UMTS-TDD (LCR) technology.

Unlicensed UMTS-TDD

In Europe, CEPT allocated the 2010-2020 MHz range for a variant of UMTS-TDD designed for unlicensed, self-provided use. Some telecom groups and jurisdictions have proposed withdrawing this service in favour of licensed UMTS-TDD, due to lack of demand, and lack of development of a UMTS TDD air interface technology suitable for deployment in this band.

Comparison with UMTS-FDD

Ordinary UMTS uses UTRA-FDD as an air interface and is known as UMTS-FDD. UMTS-FDD uses W-CDMA for multiple access and frequency division for duplexing, meaning that the up-link and down-link transmit on different frequencies. UMTS is usually transmitted on frequencies assigned for 1G, 2G, or 3G mobile telephone service in the countries of operation.

UMTS-TDD uses time division duplexing, allowing the up-link and down-link to share the same spectrum. This allows the operator to more flexibly divide the usage of available spectrum according to traffic patterns. For ordinary phone service, you would expect the up-link and down-link to carry approximately equal amounts of data (because every phone call needs a voice transmission in either direction), but Internet-oriented traffic is more frequently one-way. For example, when browsing a website, the user will send commands, which are short, to the server, but the server will send whole files, that are generally larger than those commands, in response.

UMTS-TDD tends to be allocated frequency intended for mobile/wireless Internet services rather than used on existing cellular frequencies. This is, in part, because TDD duplexing is not normally allowed on cellular, PCS/PCN, and 3G frequencies. TDD technologies open up the usage of left-over unpaired spectrum.

Europe-wide, several bands are provided either specifically for UMTS-TDD or for similar technologies. These are 1900 MHz and 1920 MHz and between 2010 MHz and 2025 MHz.

In several countries the 2500-2690 MHz band (also known as MMDS in the USA) have been used for UMTS-TDD deployments. Additionally, spectrum around the 3.5 GHz range has been allocated in some countries, notably Britain, in a technology-neutral environment. In the Czech Republic UTMS-TDD is also used in a frequency range around 872 MHz.

Deployment

UMTS-TDD has been deployed for public and/or private networks in at least nineteen countries around the world, with live systems in, amongst other countries, Australia, Czech Republic, France, Germany, Japan, New Zealand, South Africa, the UK, and the USA.

Deployments in the US thus far have been limited. It has been selected for a public safety support network used by emergency responders in New York, but outside of some experimental systems, notably one from Nextel, thus far the WiMAX standard appears to have gained greater traction as a general mobile Internet access system.

Competing Standards

A variety of Internet-access systems exist which provide broadband speed access to the net. These include WiMAX and HIPERMAN. UMTS-TDD has the advantages of being able to use an operator's existing UMTS/GSM infrastructure, should it have one, and that it includes UMTS modes optimized for circuit switching should, for example, the operator want to offer telephone service. UMTS-TDD's performance is also more consistent. However, UMTS-TDD deployers often have regulatory problems with taking advantage of some of the services UMTS compatibility provides. For example, UMTS-TDD spectrum in the UK cannot be used to provide telephone service, though the regulator OFCOM is discussing the possibility of allowing it at some point in the future. Few operators considering UMTS-TDD have existing UMTS/GSM infrastructure.

Additionally, the WiMAX and HIPERMAN systems provide significantly larger bandwidths when the mobile station is in close proximity to the tower.

Like most mobile Internet access systems, many users who might otherwise choose UMTS-TDD will find their needs covered by the ad hoc collection of unconnected Wifi access points at many restaurants and transportation hubs, and/or by Internet access already provided by their mobile phone operator. By comparison, UMTS-TDD (and systems like WiMAX) offers mobile, and more consistent, access than the former, and generally faster access than the latter.

Radio Access Network

UMTS also specifies the Universal Terrestrial Radio Access Network (UTRAN), which is composed of multiple base stations, possibly using different terrestrial air interface standards and frequency bands.

UMTS and GSM/EDGE can share a Core Network (CN), making UTRAN an alternative radio access network to GERAN (GSM/EDGE RAN), and allowing (mostly) transparent switching between the RANs according to available coverage and service needs. Because of that, UMTS's and GSM/EDGE's radio access networks are sometimes collectively referred to as UTRAN/GERAN.

UMTS networks are often combined with GSM/EDGE, the latter of which is also a part of IMT-2000.

The UE (User Equipment) interface of the RAN (Radio Access Network) primarily consists of RRC (Radio Resource Control), PDCP (Packet Data Convergence Protocol), RLC (Radio Link Control) and MAC (Media Access Control) protocols. RRC protocol handles connection establishment, measurements, radio bearer services, security and handover decisions. RLC protocol primarily divides into three Modes—Transparent Mode (TM), Unacknowledge Mode (UM), Acknowledge Mode (AM). The functionality of AM entity resembles TCP operation whereas UM operation resembles UDP operation. In TM mode, data will be sent to lower layers without adding any header to SDU of higher layers. MAC handles the scheduling of data on air interface depending on higher layer (RRC) configured parameters.

The set of properties related to data transmission is called Radio Bearer (RB). This set of properties decides the maximum allowed data in a TTI (Transmission Time Interval). RB includes RLC information and RB mapping. RB mapping decides the mapping between RB<->logical channel<->transport channel. Signaling messages are sent on Signaling Radio Bearers (SRBs) and data packets (either CS or PS) are sent on data RBs. RRC and NAS messages go on SRBs.

Security includes two procedures: integrity and ciphering. Integrity validates the resource of messages and also makes sure that no one (third/unknown party) on the radio interface has modified the messages. Ciphering ensures that no one listens to your data on the air interface. Both integrity and ciphering are applied for SRBs whereas only ciphering is applied for data RBs.

Core Network

With Mobile Application Part, UMTS uses the same core network standard as GSM/EDGE. This allows a simple migration for existing GSM operators. However, the migration path to UMTS is still costly: while much of the core infrastructure is shared with GSM, the cost of obtaining new spectrum licenses and overlaying UMTS at existing towers is high.

The CN can be connected to various backbone networks, such as the Internet or an Integrated Services Digital Network (ISDN) telephone network. UMTS (and GERAN) include the three lowest layers of OSI model. The network layer (OSI 3) includes the Radio Resource Management protocol (RRM) that manages the bearer channels between the mobile terminals and the fixed network, including the handovers.

Frequency Bands and Channel Bandwidths

UARFCN

A UARFCN (abbreviation for UTRA Absolute Radio Frequency Channel Number, where UTRA stands for UMTS Terrestrial Radio Access) is used to identify a frequency in the UMTS frequency bands.

Typically channel number is derived from the frequency in MHz through the formula Channel Number = Frequency * 5. However, this is only able to represent channels that are centered on a multiple of 200 kHz, which do not align with licensing in North America. 3GPP added several special values for the common North American channels.

Spectrum Allocation

Over 130 licenses have already been awarded to operators worldwide (as of December 2004), specifying W-CDMA radio access technology that builds on GSM. In Europe, the license process occurred at the tail end of the technology bubble, and the auction mechanisms for allocation set up in some countries resulted in some extremely high prices being paid for the original 2100 MHz licenses, notably in the UK and Germany. In Germany, bidders paid a total €50.8 billion for six licenses, two of which were subsequently abandoned and written off by their purchasers (Mobilcom and the Sonera/ Telefonica consortium). It has been suggested that these huge license fees have the character of a very large tax paid on future income expected many years down the road. In any event, the high prices paid put some European telecom operators close to bankruptcy (most notably KPN). Over the last few years some operators have written off some or all of the license costs. Between 2007 and 2009, all three Finnish carriers began to use 900 MHz UMTS in a shared arrangement with its surrounding 2G GSM base stations for rural area coverage, a trend that is expected to expand over Europe in the next 1–3 years.

The 2100 MHz band (downlink around 2100 MHz and uplink around 1900 MHz) allocated for UMTS in Europe and most of Asia is already used in North America. The 1900 MHz range is used for 2G (PCS) services, and 2100 MHz range is used for satellite communications. Regulators have, however, freed up some of the 2100 MHz range for 3G services, together with a different range around 1700 MHz for the uplink.

AT&T Wireless launched UMTS services in the United States by the end of 2004 strictly using the existing 1900 MHz spectrum allocated for 2G PCS services. Cingular acquired AT&T Wireless in 2004 and has since then launched UMTS in select US cities. Cingular renamed itself AT&T Mobility and is rolling out some cities with a UMTS network at 850 MHz to enhance its existing UMTS network at 1900 MHz and now offers subscribers a number of dual-band UMTS 850/1900 phones.

T-Mobile's rollout of UMTS in the US was originally focused on the 1700 MHz band.

However, T-Mobile has been moving users from 1700 MHz to 1900 MHz (PCS) in order to reallocate the spectrum to 4G LTE services.

In Canada, UMTS coverage is being provided on the 850 MHz and 1900 MHz bands on the Rogers and Bell-Telus networks. Bell and Telus share the network. Recently, new providers Wind Mobile, Mobilicity and Videotron have begun operations in the 1700 MHz band.

In 2008, Australian telco Telstra replaced its existing CDMA network with a national UMTS-based 3G network, branded as NextG, operating in the 850 MHz band. Telstra currently provides UMTS service on this network, and also on the 2100 MHz UMTS network, through a co-ownership of the owning and administrating company 3GIS. This company is also co-owned by Hutchison 3G Australia, and this is the primary network used by their customers. Optus is currently rolling out a 3G network operating on the 2100 MHz band in cities and most large towns, and the 900 MHz band in regional areas. Vodafone is also building a 3G network using the 900 MHz band.

In India, BSNL has started its 3G services since October 2009, beginning with the larger cities and then expanding over to smaller cities. The 850 MHz and 900 MHz bands provide greater coverage compared to equivalent 1700/1900/2100 MHz networks, and are best suited to regional areas where greater distances separate base station and subscriber.

Carriers in South America are now also rolling out 850 MHz networks.

Interoperability and Global Roaming

UMTS phones (and data cards) are highly portable—they have been designed to roam easily onto other UMTS networks (if the providers have roaming agreements in place). In addition, almost all UMTS phones are UMTS/GSM dual-mode devices, so if a UMTS phone travels outside of UMTS coverage during a call the call may be transparently handed off to available GSM coverage. Roaming charges are usually significantly higher than regular usage charges.

Most UMTS licensees consider ubiquitous, transparent global roaming an important issue. To enable a high degree of interoperability, UMTS phones usually support several different frequencies in addition to their GSM fallback. Different countries support different UMTS frequency bands – Europe initially used 2100 MHz while the most carriers in the USA use 850 MHz and 1900 MHz. T-Mobile has launched a network in the US operating at 1700 MHz (uplink) /2100 MHz (downlink), and these bands also have been adopted elsewhere in the US and in Canada and Latin America. A UMTS phone and network must support a common frequency to work together. Because of the frequencies used, early models of UMTS phones designated for the United States will likely not be operable elsewhere and vice versa. There are now 11 different frequency combinations used around the world—including frequencies formerly used solely for 2G services.

UMTS phones can use a Universal Subscriber Identity Module, USIM (based on GSM's SIM) and also work (including UMTS services) with GSM SIM cards. This is a global standard of identification, and enables a network to identify and authenticate the (U) SIM in the phone. Roaming agreements between networks allow for calls to a customer to be redirected to them while roaming and determine the services (and prices) available to the user. In addition to user subscriber information and authentication information, the (U)SIM provides storage space for phone book contact. Handsets can store their data on their own memory or on the (U)SIM card (which is usually more limited in its phone book contact information). A (U)SIM can be moved to another UMTS or GSM phone, and the phone will take on the user details of the (U)SIM, meaning it is the (U)SIM (not the phone) which determines the phone number of the phone and the billing for calls made from the phone.

Japan was the first country to adopt 3G technologies, and since they had not used GSM previously they had no need to build GSM compatibility into their handsets and their 3G handsets were smaller than those available elsewhere. In 2002, NTT DoCoMo's FOMA 3G network was the first commercial UMTS network—using a pre-release specification, it was initially incompatible with the UMTS standard at the radio level but used standard USIM cards, meaning USIM card based roaming was possible (transferring the USIM card into a UMTS or GSM phone when travelling). Both NTT DoCoMo and SoftBank Mobile (which launched 3G in December 2002) now use standard UMTS.

Handsets and Modems

The Nokia 6650, an early (2003) UMTS handset

All of the major 2G phone manufacturers (that are still in business) are now manufacturers of 3G phones. The early 3G handsets and modems were specific to the frequencies required in their country, which meant they could only roam to other countries on the same 3G frequency (though they can fall back to the older GSM standard). Canada and USA have a common share of frequencies, as do most European countries. The article UMTS frequency bands is an overview of UMTS network frequencies around the world.

Using a cellular router, PCMCIA or USB card, customers are able to access 3G broadband services, regardless of their choice of computer (such as a tablet PC or a PDA). Some software installs itself from the modem, so that in some cases absolutely no knowledge of technology is required to get online in moments. Using a phone that supports 3G and Bluetooth 2.0, multiple Bluetooth-capable laptops can be connected to the Internet. Some smartphones can also act as a mobile WLAN access point.

There are very few 3G phones or modems available supporting all 3G frequencies (U MTS850/900/1700/1900/2100 MHz). Nokia has recently released a range of phones that have Pentaband 3G coverage, including the N8 and E7. Many other phones are offering more than one band which still enables extensive roaming. For example, Apple's iPhone 4 contains a quadband chipset operating on 850/900/1900/2100 MHz, allowing usage in the majority of countries where UMTS-FDD is deployed.

Other Competing Standards

The main competitor to UMTS is CDMA2000 (IMT-MC), which is developed by the 3GPP2. Unlike UMTS, CDMA2000 is an evolutionary upgrade to an existing 2G standard, cdma-One, and is able to operate within the same frequency allocations. This and CDMA2000's narrower bandwidth requirements make it easier to deploy in existing spectra. In some, but not all, cases, existing GSM operators only have enough spectrum to implement either UMTS or GSM, not both. For example, in the US D, E, and F PCS spectrum blocks, the amount of spectrum available is 5 MHz in each direction. A standard UMTS system would saturate that spectrum. Where CDMA2000 is deployed, it usually co-exists with UMTS. In many markets however, the co-existence issue is of little relevance, as legislative hurdles exist to co-deploying two standards in the same licensed slice of spectrum.

Another competitor to UMTS is EDGE (IMT-SC), which is an evolutionary upgrade to the 2G GSM system, leveraging existing GSM spectrums. It is also much easier, quicker, and considerably cheaper for wireless carriers to "bolt-on" EDGE functionality by upgrading their existing GSM transmission hardware to support EDGE rather than having to install almost all brand-new equipment to deliver UMTS. However, being developed by 3GPP just as UMTS, EDGE is not a true competitor. Instead, it is used as a temporary solution preceding UMTS roll-out or as a complement for rural areas. This is facilitated by the fact that GSM/EDGE and UMTS specification are jointly developed and rely on the same core network, allowing dual-mode operation including vertical handovers.

China's TD-SCDMA standard is often seen as a competitor, too. TD-SCDMA has been added to UMTS' Release 4 as UTRA-TDD 1.28 Mcps Low Chip Rate (UTRA-TDD LCR). Unlike TD-CDMA (UTRA-TDD 3.84 Mcps High Chip Rate, UTRA-TDD HCR) which complements W-CDMA (UTRA-FDD), it is suitable for both micro and macro cells. However, the lack of vendors' support is preventing it from being a real competitor.

While DECT is technically capable of competing with UMTS and other cellular networks in densely populated, urban areas, it has only been deployed for domestic cordless phones and private in-house networks.

All of these competitors have been accepted by ITU as part of the IMT-2000 family of 3G standards, along with UMTS-FDD.

On the Internet access side, competing systems include WiMAX and Flash-OFDM.

Migrating from GSM/GPRS to UMTS

From a GSM/GPRS network, the following network elements can be reused:

- Home Location Register (HLR)
- Visitor Location Register (VLR)
- Equipment Identity Register (EIR)
- Mobile Switching Center (MSC) (vendor dependent)
- Authentication Center (AUC)
- Serving GPRS Support Node (SGSN) (vendor dependent)
- Gateway GPRS Support Node (GGSN)

From a GSM/GPRS communication radio network, the following elements cannot be reused:

- Base station controller (BSC)
- Base transceiver station (BTS)

They can remain in the network and be used in dual network operation where 2G and 3G networks co-exist while network migration and new 3G terminals become available for use in the network.

The UMTS network introduces new network elements that function as specified by 3GPP:

- Node B (base transceiver station)
- Radio Network Controller (RNC)
- Media Gateway (MGW)

The functionality of MSC and SGSN changes when going to UMTS. In a GSM system the MSC handles all the circuit switched operations like connecting A- and B-subscriber through the network. SGSN handles all the packet switched operations and transfers all the data in the network. In UMTS the Media gateway (MGW) take care of all data transfer in both circuit and packet switched networks. MSC and SGSN control MGW operations. The nodes are renamed to MSC-server and GSN-server.

Problems and Issues

Some countries, including the United States, have allocated spectrum differently from the ITU recommendations, so that the standard bands most commonly used for UMTS (UMTS-2100) have not been available. In those countries, alternative bands are used, preventing the interoperability of existing UMTS-2100 equipment, and requiring the design and manufacture of different equipment for the use in these markets. As is the case with GSM900 today, standard UMTS 2100 MHz equipment will not work in those markets. However, it appears as though UMTS is not suffering as much from handset band compatibility issues as GSM did, as many UMTS handsets are multi-band in both UMTS and GSM modes. Penta-band (850, 900, 1700 / 2100, and 1900 MHz bands), quad-band GSM (850, 900, 1800, and 1900 MHz bands) and tri-band UMTS (850, 1900, and 2100 MHz bands) handsets are becoming more commonplace.

In its early days, UMTS had problems in many countries: Overweight handsets with poor battery life were first to arrive on a market highly sensitive to weight and form factor. The Motorola A830, a debut handset on Hutchison's 3 network, weighed more than 200 grams and even featured a detachable camera to reduce handset weight. Another significant issue involved call reliability, related to problems with handover from UMTS to GSM. Customers found their connections being dropped as handovers were possible only in one direction (UMTS → GSM), with the handset only changing back to UMTS after hanging up. In most networks around the world this is no longer an issue.

Compared to GSM, UMTS networks initially required a higher base station density. For fully-fledged UMTS incorporating video on demand features, one base station needed to be set up every 1–1.5 km (0.62–0.93 mi). This was the case when only the 2100 MHz band was being used, however with the growing use of lower-frequency bands (such as 850 and 900 MHz) this is no longer so. This has led to increasing rollout of the lower-band networks by operators since 2006.

Even with current technologies and low-band UMTS, telephony and data over UMTS requires more power than on comparable GSM networks. Apple Inc. cited UMTS power consumption as the reason that the first generation iPhone only supported EDGE. Their release of the iPhone 3G quotes talk time on UMTS as half that available when the handset is set to use GSM. Other manufacturers indicate different battery lifetime for UMTS mode compared to GSM mode as well. As battery and network technology improve, this issue is diminishing.

Security Issues

As early as 2008 it was known that carrier networks can be used to surreptitiously gather user location information. In August 2014, the Washington Post reported on widespread marketing of surveillance systems using Signalling System No. 7 (SS7) protocols to locate callers anywhere in the world.

In December 2014, news broke that SS7's very own functions can be repurposed for surveillance, because of its lax security, in order to listen to calls in real time or to record encrypted calls and texts for later decryption, or to defraud users and cellular carriers.

The German Telekom and Vodafone declared the same day that they had fixed gaps in their networks, but that the problem is global and can only be fixed with a telecommunication system-wide solution.

CDMA2000

Huawei CDMA2000 EVDO USB wireless modem

CDMA2000 (also known as C2K or IMT MultiCarrier (IMTMC)) is a family of 3G mobile technology standards for sending voice, data, and signaling data between mobile phones and cell sites. It is developed by 3GPP2 as a backwards-compatible successor to second-generation cdmaOne (IS-95) set of standards and used especially in North America and South Korea.

CDMA2000 compares to UMTS, a competing set of 3G standards, which is developed by 3GPP and used in Europe, Japan, and China.

The name CDMA2000 denotes a family of standards that represent the successive, evolutionary stages of the underlying technology. These are:

- Voice: CDMA2000 1xRTT, 1X Advanced

- Data: CDMA2000 1xEV-DO (Evolution-Data Optimized): Release 0, Revision A, Revision B, Ultra Mobile Broadband (UMB)

All are approved radio interfaces for the ITU's IMT-2000. In the United States, CDMA2000 is a registered trademark of the Telecommunications Industry Association (TIA-USA).

1X

CDMA2000 1X (IS-2000), also known as 1x and 1xRTT, is the core CDMA2000 wireless air interface standard. The designation "1x", meaning *1 times Radio Transmission Technology*, indicates the same radio frequency (RF) bandwidth as IS-95: a duplex pair of 1.25 MHz radio channels. 1xRTT almost doubles the capacity of IS-95 by adding 64 more traffic channels to the forward link, orthogonal to (in quadrature with) the original set of 64. The 1X standard supports packet data speeds of up to 153 kbit/s with real world data transmission averaging 80–100 kbit/s in most commercial applications. IMT-2000 also made changes to the data link layer for greater use of data services, including medium and link access control protocols and QoS. The IS-95 data link layer only provided "best efforts delivery" for data and circuit switched channel for voice (i.e., a voice frame once every 20 ms).

1xEV-DO

BlackBerry smartphone displaying '1XEV' as the service status in the upper right corner.

CDMA2000 1xEV-DO (Evolution-Data Optimized), often abbreviated as EV-DO or EV, is a telecommunications standard for the wireless transmission of data through radio signals, typically for broadband Internet access. It uses multiplexing techniques including code division multiple access (CDMA) as well as time division multiple access (TDMA) to maximize both individual user's throughput and the overall system throughput. It is standardized by 3rd Generation Partnership Project 2 (3GPP2) as part of the CDMA2000 family of standards and has been adopted by many mobile phone service providers around the world – particularly those previously employing CDMA networks.

1X Advanced

1X Advanced(Rev.E) is the evolution of CDMA2000 1X. It provides up to four times the capacity and 70% more coverage compared to 1X.

Networks

The CDMA Development Group states that, as of April 2014, there are 314 operators in 118 countries offering CDMA2000 1X and/or 1xEV-DO service.

History

The intended 4G successor to CDMA2000 was UMB (Ultra Mobile Broadband); however, in November 2008, Qualcomm announced it was ending development of the technology, favoring LTE instead.

LTE (Telecommunication)

Adoption of LTE technology

▎Countries and regions with commercial LTE service
▎Countries and regions with commercial LTE network deployment on-going or planned
▎Countries and regions with LTE trial systems (pre-commitment)

LTE signal indicator in Android

Long-Term Evolution (LTE) is a standard for high-speed wireless communication for mobile phones and data terminals. It is based on the GSM/EDGE and UMTS/HSPA network technologies, increasing the capacity and speed using a different radio interface together with core network improvements. The standard is developed by the 3GPP (3rd Generation Partnership Project) and is specified in its Release 8 document series, with minor enhancements described in Release 9. Y LTE is the upgrade same for carriers with both GSM/UMTS networks and CDMA2000 networks. The different LTE frequencies and bands used in different countries will mean that only multi-band phones will be able to use LTE in all countries where it is supported.

LTE is commonly marketed as 4G LTE, but it does not meet the technical criteria of a 4G wireless service, as specified in the 3GPP Release 8 and 9 document series, for LTE Advanced. The requirements were originally set forth by the ITU-R organization in the IMT Advanced specification. However, due to marketing pressures and the significant advancements that WiMAX, Evolved High Speed Packet Access and LTE bring to the original 3G technologies, ITU later decided that LTE together with the aforementioned technologies can be called 4G technologies. The LTE Advanced standard formally satisfies the ITU-R requirements to be considered IMT-Advanced. To differentiate LTE Advanced and WiMAX-Advanced from current 4G technologies, ITU has defined them as "True 4G".

Overview

Telia-branded Samsung LTE modem

LTE stands for Long Term Evolution and is a registered trademark owned by ETSI (European Telecommunications Standards Institute) for the wireless data communications technology and a development of the GSM/UMTS standards. However, other nations and companies do play an active role in the LTE project. The goal of LTE was to increase the capacity and speed of wireless data networks using new DSP (digital signal processing) techniques and modulations that were developed around the turn of the millennium. A further goal was the redesign and simplification of the network architecture to an IP-based system with significantly reduced transfer latency compared to the 3G architecture. The LTE wireless interface is incompatible with 2G and 3G networks, so that it must be operated on a separate radio spectrum.

HTC ThunderBolt, the second commercially available LTE smartphone

LTE was first proposed by NTT DoCoMo of Japan in 2004, and studies on the new standard officially commenced in 2005. In May 2007, the LTE/SAE Trial Initiative (LSTI) alliance was founded as a global collaboration between vendors and operators with the goal of verifying and promoting the new standard in order to ensure the global introduction of the technology as quickly as possible. The LTE standard was finalized in December 2008, and the first publicly available LTE service was launched by TeliaSonera in Oslo and Stockholm on December 14, 2009 as a data connection with a USB modem. The LTE services were launched by major North American carriers as well, with the Samsung SCH-r900 being the world's first LTE Mobile phone starting on September 21, 2010 and Samsung Galaxy Indulge being the world's first LTE smartphone starting on February 10, 2011 both offered by MetroPCS and HTC ThunderBolt offered by Verizon starting on March 17 being the second LTE smartphone to be sold commercially. In Canada, Rogers Wireless was the first to launch LTE network on July 7, 2011 offering the Sierra Wireless AirCard® 313U USB mobile broadband modem, known as the "LTE Rocket™ stick" then followed closely by mobile devices from both HTC and Samsung. Initially, CDMA operators planned to upgrade to rival standards called UMB and WiMAX, but all the major CDMA operators (such as Verizon, Sprint and MetroPCS in the United States, Bell and Telus in Canada, au by KDDI in Japan, SK Telecom in South Korea and China Telecom/China Unicom in China) have announced that they intend to migrate to LTE after all. The evolution of LTE is LTE Advanced, which was standardized in March 2011. Services are expected to commence in 2013.

The LTE specification provides downlink peak rates of 300 Mbit/s, uplink peak rates of 75 Mbit/s and QoS provisions permitting a transfer latency of less than 5 ms in the radio access network. LTE has the ability to manage fast-moving mobiles and supports multi-cast and broadcast streams. LTE supports scalable carrier bandwidths, from 1.4 MHz to 20 MHz and supports both frequency division duplexing (FDD) and time-division duplexing (TDD). The IP-based network architecture, called the Evolved Packet Core (EPC) designed to replace the GPRS Core Network, supports seamless handovers for both voice and data to cell towers with older network technology such as GSM, UMTS and CDMA2000. The simpler architecture results in lower operating costs (for example, each E-UTRA cell will support up to four times the data and voice capacity supported by HSPA).

History

3GPP Standard Development Timeline

- In 2004, NTT DoCoMo of Japan proposes LTE as the international standard.

- In September 2006, Siemens Networks (today Nokia Networks) showed in collaboration with Nomor Research the first live emulation of an LTE network to the media and investors. As live applications two users streaming an HDTV video in the downlink and playing an interactive game in the uplink have been demonstrated.

- In February 2007, Ericsson demonstrated for the first time in the world LTE with bit rates up to 144 Mbit/s

- In September 2007, NTT docomo demonstrated LTE data rates of 200 Mbit/s with power level below 100 mW during the test.

- In November 2007, Infineon presented the world's first RF transceiver named SMARTi LTE supporting LTE functionality in a single-chip RF silicon processed in CMOS

- In early 2008, LTE test equipment began shipping from several vendors and, at the Mobile World Congress 2008 in Barcelona, Ericsson demonstrated the world's first end-to-end mobile call enabled by LTE on a small handheld device. Motorola demonstrated an LTE RAN standard compliant eNodeB and LTE chipset at the same event.

- At the February 2008 Mobile World Congress:

  - Motorola demonstrated how LTE can accelerate the delivery of personal media experience with HD video demo streaming, HD video blogging, Online gaming and VoIP over LTE running a RAN standard compliant LTE network & LTE chipset.

  - Ericsson EMP (now ST-Ericsson) demonstrated the world's first end-to-end LTE call on handheld Ericsson demonstrated LTE FDD and TDD mode on the same base station platform.

- Freescale Semiconductor demonstrated streaming HD video with peak data rates of 96 Mbit/s downlink and 86 Mbit/s uplink.

- NXP Semiconductors (now a part of ST-Ericsson) demonstrated a multi-mode LTE modem as the basis for a software-defined radio system for use in cellphones.

- picoChip and Mimoon demonstrated a base station reference design. This runs on a common hardware platform (multi-mode / software-defined radio) with their WiMAX architecture.

- In April 2008, Motorola demonstrated the first EV-DO to LTE hand-off – handing over a streaming video from LTE to a commercial EV-DO network and back to LTE.

- In April 2008, LG Electronics and Nortel demonstrated LTE data rates of 50 Mbit/s while travelling at 110 km/h.

- In November 2008, Motorola demonstrated industry first over-the-air LTE session in 700 MHz spectrum.

- Researchers at Nokia Siemens Networks and Heinrich Hertz Institut have demonstrated LTE with 100 Mbit/s Uplink transfer speeds.

- At the February 2009 Mobile World Congress:

  - Infineon demonstrated a single-chip 65 nm CMOS RF transceiver providing 2G/3G/LTE functionality

  - Launch of ng Connect program, a multi-industry consortium founded by Alcatel-Lucent to identify and develop wireless broadband applications.

  - Motorola provided LTE drive tour on the streets of Barcelona to demonstrate LTE system performance in a real-life metropolitan RF environment

- In July 2009, Nujira demonstrated efficiencies of more than 60% for an 880 MHz LTE Power Amplifier

- In August 2009, Nortel and LG Electronics demonstrated the first successful handoff between CDMA and LTE networks in a standards-compliant manner

- In August 2009, Alcatel-Lucent receives FCC certification for LTE base stations for the 700 MHz spectrum band.

- In September 2009, Nokia Siemens Networks demonstrated world's first LTE call on standards-compliant commercial software.

- In October 2009, Ericsson and Samsung demonstrated interoperability between the first ever commercial LTE device and the live network in Stockholm, Sweden.

- In October 2009, Alcatel-Lucent's Bell Labs, Deutsche Telekom Laboratories, the Fraunhofer Heinrich-Hertz Institut and antenna supplier Kathrein conducted live field tests of a technology called Coordinated Multipoint Transmission (CoMP) aimed at increasing the data transmission speeds of Long Term Evolution (LTE) and 3G networks.

- In November 2009, Alcatel-Lucent completed first live LTE call using 800 MHz spectrum band set aside as part of the European Digital Dividend (EDD).

- In November 2009, Nokia Siemens Networks and LG completed first end-to-end interoperability testing of LTE.

- On December 14, 2009, the first commercial LTE deployment was in the Scandinavian capitals Stockholm and Oslo by the Swedish-Finnish network operator TeliaSonera and its Norwegian brandname NetCom (Norway). TeliaSonera incorrectly branded the network "4G". The modem devices on offer were manufactured by Samsung (dongle GT-B3710), and the network infrastructure with SingleRAN technology created by Huawei (in Oslo) and Ericsson (in Stockholm). TeliaSonera plans to roll out nationwide LTE across Sweden, Norway and Finland. TeliaSonera used spectral bandwidth of 10 MHz (out of the maximum 20 MHz), and Single-Input and Single-Output transmission. The deployment should have provided a physical layer net bitrates of up to 50 Mbit/s downlink and 25 Mbit/s in the uplink. Introductory tests showed a TCP goodput of 42.8 Mbit/s downlink and 5.3 Mbit/s uplink in Stockholm.

- In December 2009, ST-Ericsson and Ericsson first to achieve LTE and HSPA mobility with a multimode device.

- In January 2010, Alcatel-Lucent and LG complete a live handoff of an end-to-end data call between Long Term Evolution (LTE) and CDMA networks.

- In February 2010, Nokia Siemens Networks and Movistar test the LTE in Mobile World Congress 2010 in Barcelona, Spain, with both indoor and outdoor demonstrations.

- In May 2010, Mobile TeleSystems (MTS) and Huawei showed an indoor LTE network at "Sviaz-Expocomm 2010" in Moscow, Russia. MTS expects to start a trial LTE service in Moscow by the beginning of 2011. Earlier, MTS has received a license to build an LTE network in Uzbekistan, and intends to commence a test LTE network in Ukraine in partnership with Alcatel-Lucent.

- At the Shanghai Expo 2010 in May 2010, Motorola demonstrated a live LTE in conjunction with China Mobile. This included video streams and a drive test system using TD-LTE.

- As of 12/10/2010, DirecTV has teamed up with Verizon Wireless for a test of high-speed Long Term Evolution (LTE) wireless technology in a few homes in

Pennsylvania, designed to deliver an integrated Internet and TV bundle. Verizon Wireless said it launched LTE wireless services (for data, no voice) in 38 markets where more than 110 million Americans live on Sunday, Dec. 5.

- On May 6, 2011, Sri Lanka Telecom Mobitel successfully demonstrated 4G LTE for the first time in South Asia, achieving a data rate of 96 Mbit/s in Sri Lanka.

Carrier Adoption Timeline

Most carriers supporting GSM or HSUPA networks can be expected to upgrade their networks to LTE at some stage. A complete list of commercial contracts can be found at:

- August 2009: Telefónica selected six countries to field-test LTE in the succeeding months: Spain, the United Kingdom, Germany and the Czech Republic in Europe, and Brazil and Argentina in Latin America.

- On November 24, 2009: Telecom Italia announced the first outdoor pre-commercial experimentation in the world, deployed in Torino and totally integrated into the 2G/3G network currently in service.

- On December 14, 2009, the world's first publicly available LTE service was opened by TeliaSonera in the two Scandinavian capitals Stockholm and Oslo.

- On May 28, 2010, Russian operator Scartel announced the launch of an LTE network in Kazan by the end of the 2010.

- On October 6, 2010, Canadian provider Rogers Communications Inc announced that Ottawa, Canada's national capital, will be the site of LTE trials. Rogers said it will expand on this testing and move to a comprehensive technical trial of LTE on both low- and high-band frequencies across the Ottawa area.

- On May 6, 2011, Sri Lanka Telecom Mobitel successfully demonstrated 4G LTE for the first time in South Asia, achieving a data rate of 96 Mbit/s in Sri Lanka.

- On May 7, 2011, Sri Lankan Mobile Operator Dialog Axiata PLC switched on the first pilot 4G LTE Network in South Asia with vendor partner Huawei and demonstrated a download data speed up to 127 Mbit/s.

- On February 9, 2012, Telus Mobility launched their LTE service initial in metropolitan areas include Vancouver, Calgary, Edmonton, Toronto and the Greater Toronto Area, Kitchener, Waterloo, Hamilton, Guelph, Belleville, Ottawa, Montreal, Québec City, Halifax and Yellowknife.

- Telus Mobility has announced that it will adopt LTE as its 4G wireless standard.

- Cox Communications has its first tower for wireless LTE network build-out. Wireless services launched in late 2009.

Below is a list of countries by 4G LTE penetration as measured by OpenSignal.com in 2015.

LTE-TDD

Long-Term Evolution Time-Division Duplex (LTE-TDD), also referred to as TDD LTE, is a 4G telecommunications technology and standard co-developed by an international coalition of companies, including China Mobile, Datang Telecom, Huawei, ZTE, Nokia Solutions and Networks, Qualcomm, Samsung, and ST-Ericsson. It is one of the two mobile data transmission technologies of the Long-Term Evolution (LTE) technology standard, the other being Frequency-Division Long-Term Evolution (LTE-FDD). While some companies refer to LTE TDD as "TD-LTE", there is no reference to that acronym anywhere in the 3GPP specifications.

There are two major differences between LTE-TDD and LTE-FDD: how data is uploaded and downloaded, and what frequency spectra the networks are deployed in. While LTE-FDD uses paired frequencies to upload and download data, LTE-TDD uses a single frequency, alternating between uploading and downloading data through time. The ratio between uploads and downloads on a LTE-TDD network can be changed dynamically, depending on whether more data needs to be sent or received. LTE-TDD and LTE-FDD also operate on different frequency bands, with LTE-TDD working better at higher frequencies, and LTE-FDD working better at lower frequencies. Frequencies used for LTE-TDD range from 1850 MHz to 3800 MHz, with several different bands being used. The LTE-TDD spectrum is generally cheaper to access, and has less traffic. Further, the bands for LTE-TDD overlap with those used for WiMAX, which can easily be upgraded to support LTE-TDD.

Despite the differences in how the two types of LTE handle data transmission, LTE-TDD and LTE-FDD share 90 percent of their core technology, making it possible for the same chipsets and networks to use both versions of LTE. A number of companies produce dual-mode chips or mobile devices, including Samsung and Qualcomm, while operators China Mobile Hong Kong Company Limited and Hi3G Access have developed dual-mode networks in China and Sweden, respectively.

History

The creation of LTE-TDD involved a coalition of international companies that worked to develop and test the technology. China Mobile was an early proponent of LTE-TDD, along with other companies like Datang Telecom and Huawei, which worked to deploy LTE-TDD networks, and later developed technology allowing LTE-TDD equipment to operate in white spaces—frequency spectra between broadcast TV stations. Intel also participated in the development, setting up a LTE-TDD interoperability lab with Huawei in China, as well as ST-Ericsson, Nokia, and Nokia Siemens (now Nokia Solutions and Networks), which developed LTE-TDD base stations that increased capacity by

80 percent and coverage by 40 percent. Qualcomm also participated, developing the world's first multi-mode chip, combining both LTE-TDD and LTE-FDD, along with HSPA and EV-DO. Accelleran, a Belgian company, has also worked to build small cells for LTE-TDD networks.

Trials of LTE-TDD technology began as early as 2010, with Reliance Industries and Ericsson India conducting field tests of LTE-TDD in India, achieving 80 megabit-per second download speeds and 20 megabit-per-second upload speeds. By 2011, China Mobile began trials of the technology in six cities.

Although initially seen as a technology utilized by only a few countries, including China and India, by 2011 international interest in LTE-TDD had expanded, especially in Asia, in part due to LTE-TDD 's lower cost of deployment compared to LTE-FDD. By the middle of that year, 26 networks around the world were conducting trials of the technology. The Global LTE-TDD Initiative (GTI) was also started in 2011, with founding partners China Mobile, Bharti Airtel, SoftBank Mobile, Vodafone, Clearwire, Aero2 and E-Plus. In September 2011, Huawei announced it would partner with Polish mobile provider Aero2 to develop a combined LTE TDD and FDD network in Poland, and by April 2012, ZTE Corporation had worked to deploy trial or commercial LTE-TDD networks for 33 operators in 19 countries. In late 2012, Qualcomm worked extensively to deploy a commercial LTE-TDD network in India, and partnered with Bharti Airtel and Huawei to develop the first multi-mode LTE-TDD smartphone for India.

In Japan, SoftBank Mobile launched LTE-TDD services in February 2012 under the name Advanced eXtended Global Platform (AXGP), and marketed as SoftBank 4G (ja). The AXGP band was previously used for Willcom's PHS service, and after PHS was discontinued in 2010 the PHS band was re-purposed for AXGP service.

In the U.S., Clearwire planned to implement LTE-TDD, with chip-maker Qualcomm agreeing to support Clearwire's frequencies on its multi-mode LTE chipsets. With Sprint's acquisition of Clearwire in 2013, the carrier began using these frequencies for LTE service on networks built by Samsung, Alcatel-Lucent, and Nokia.

As of March 2013, 156 commercial 4G LTE networks existed, including 142 LTE-FDD networks and 14 LTE-TDD networks. As of November 2013, the South Korean government planned to allow a fourth wireless carrier in 2014, which would provide LTE-TDD services, and in December 2013, LTE-TDD licenses were granted to China's three mobile operators, allowing commercial deployment of 4G LTE services.

In January 2014, Nokia Solutions and Networks indicated that it had completed a series of tests of voice over LTE (VoLTE) calls on China Mobile's TD-LTE network. The next month, Nokia Solutions and Networks and Sprint announced that they had demonstrated throughput speeds of 2.6 gigabits per second throughput using a LTE-TDD network, surpassing the previous record of 1.6 gigbits per second.

LTE Direct

A new LTE protocol named LTE Direct works as an innovative device-to-device technology enabling the discovery of thousands of devices in the proximity of approximately 500 meters. Pioneered by Qualcomm, the company has been leading the standardization of this new technology along with other 3GPP participants. LTE Direct offers several advantages over existing proximity solutions including but not limited to Wi-Fi or Bluetooth. One of the most popular use cases for this technology was developed by a New York City based company called Compass.to. The core feature of proximal discovery among devices included a targeted discount voucher to a nearby device which matched specific interests. The Compass.to use case was featured at global conferences and events such as CES 2015, MWC 2015, and said to be extended to many other scenarios including film festivals, theme parks and sporting events. "You can think of LTE Direct as a sixth sense that is always aware of the environment around you," said Mahesh Makhijani, technical marketing director at Qualcomm, at a session on the technology. Additionally, the protocol offers less battery drainage and extended range when compared to other proximity solutions.

Features

Much of the LTE standard addresses the upgrading of 3G UMTS to what will eventually be 4G mobile communications technology. A large amount of the work is aimed at simplifying the architecture of the system, as it transitions from the existing UMTS circuit + packet switching combined network, to an all-IP flat architecture system. E-UTRA is the air interface of LTE. Its main features are:

- Peak download rates up to 299.6 Mbit/s and upload rates up to 75.4 Mbit/s depending on the user equipment category (with 4×4 antennas using 20 MHz of spectrum). Five different terminal classes have been defined from a voice centric class up to a high end terminal that supports the peak data rates. All terminals will be able to process 20 MHz bandwidth.

- Low data transfer latencies (sub-5 ms latency for small IP packets in optimal conditions), lower latencies for handover and connection setup time than with previous radio access technologies.

- Improved support for mobility, exemplified by support for terminals moving at up to 350 km/h (220 mph) or 500 km/h (310 mph) depending on the frequency band.

- Orthogonal frequency-division multiple access for the downlink, Single-carrier FDMA for the uplink to conserve power.

- Support for both FDD and TDD communication systems as well as half-duplex FDD with the same radio access technology.

- Support for all frequency bands currently used by IMT systems by ITU-R.

- Increased spectrum flexibility: 1.4 MHz, 3 MHz, 5 MHz, 10 MHz, 15 MHz and 20 MHz wide cells are standardized. (W-CDMA has no option for other than 5 MHz slices, leading to some problems rolling-out in countries where 5 MHz is a commonly allocated width of spectrum so would frequently already be in use with legacy standards such as 2G GSM and cdmaOne.)

- Support for cell sizes from tens of metres radius (femto and picocells) up to 100 km (62 miles) radius macrocells. In the lower frequency bands to be used in rural areas, 5 km (3.1 miles) is the optimal cell size, 30 km (19 miles) having reasonable performance, and up to 100 km cell sizes supported with acceptable performance. In city and urban areas, higher frequency bands (such as 2.6 GHz in EU) are used to support high speed mobile broadband. In this case, cell sizes may be 1 km (0.62 miles) or even less.

- Supports at least 200 active data clients in every 5 MHz cell.

- Simplified architecture: The network side of E-UTRAN is composed only of eNode Bs.

- Support for inter-operation and co-existence with legacy standards (e.g., GSM/EDGE, UMTS and CDMA2000). Users can start a call or transfer of data in an area using an LTE standard, and, should coverage be unavailable, continue the operation without any action on their part using GSM/GPRS or W-CDMA-based UMTS or even 3GPP2 networks such as cdmaOne or CDMA2000.

- Packet switched radio interface.

- Support for MBSFN (Multicast-broadcast single-frequency network). This feature can deliver services such as Mobile TV using the LTE infrastructure, and is a competitor for DVB-H-based TV broadcast.

Voice Calls

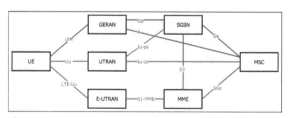

cs domLTE CSFB to GSM/UMTS network interconnects

The LTE standard supports only packet switching with its all-IP network. Voice calls in GSM, UMTS and CDMA2000 are circuit switched, so with the adoption of LTE, carriers will have to re-engineer their voice call network. Three different approaches sprang up:

Voice over LTE (VoLTE)

Circuit-switched fallback (CSFB)

> In this approach, LTE just provides data services, and when a voice call is to be initiated or received, it will fall back to the circuit-switched domain. When using this solution, operators just need to upgrade the MSC instead of deploying the IMS, and therefore, can provide services quickly. However, the disadvantage is longer call setup delay.

Simultaneous voice and LTE (SVLTE)

> In this approach, the handset works simultaneously in the LTE and circuit switched modes, with the LTE mode providing data services and the circuit switched mode providing the voice service. This is a solution solely based on the handset, which does not have special requirements on the network and does not require the deployment of IMS either. The disadvantage of this solution is that the phone can become expensive with high power consumption.

One additional approach which is not initiated by operators is the usage of over-the-top content (OTT) services, using applications like Skype and Google Talk to provide LTE voice service.

Most major backers of LTE preferred and promoted VoLTE from the beginning. The lack of software support in initial LTE devices as well as core network devices however led to a number of carriers promoting VoLGA (Voice over LTE Generic Access) as an interim solution. The idea was to use the same principles as GAN (Generic Access Network, also known as UMA or Unlicensed Mobile Access), which defines the protocols through which a mobile handset can perform voice calls over a customer's private Internet connection, usually over wireless LAN. VoLGA however never gained much support, because VoLTE (IMS) promises much more flexible services, albeit at the cost of having to upgrade the entire voice call infrastructure. VoLTE will also require Single Radio Voice Call Continuity (SRVCC) in order to be able to smoothly perform a handover to a 3G network in case of poor LTE signal quality.

While the industry has seemingly standardized on VoLTE for the future, the demand for voice calls today has led LTE carriers to introduce CSFB as a stopgap measure. When placing or receiving a voice call, LTE handsets will fall back to old 2G or 3G networks for the duration of the call.

Enhanced Voice Quality

To ensure compatibility, 3GPP demands at least AMR-NB codec (narrow band), but the recommended speech codec for VoLTE is Adaptive Multi-Rate Wideband, also known as HD Voice. This codec is mandated in 3GPP networks that support 16 kHz sampling.

Fraunhofer IIS has proposed and demonstrated "Full-HD Voice", an implementation of the AAC-ELD (Advanced Audio Coding – Enhanced Low Delay) codec for LTE handsets. Where previous cell phone voice codecs only supported frequencies up to 3.5 kHz and upcoming wideband audio services branded as *HD Voice* up to 7 kHz, Full-HD Voice supports the entire bandwidth range from 20 Hz to 20 kHz. For end-to-end Full-HD Voice calls to succeed however, both the caller and recipient's handsets as well as networks have to support the feature.

Frequency Bands

The LTE standard covers a range of many different bands, each of which is designated by both a frequency and a band number. In North America, 700, 750, 800, 850, 1900, 1700/2100 (AWS), 2300 (WCS) 2500 and 2600 MHz (Rogers Communications, Bell Canada) are used (bands 2, 4, 5, 7, 12, 13, 17, 25, 26, 30, 41); 2500 MHz in South America; 700, 800, 900, 1800, 2600 MHz in Europe (bands 3, 7, 20); 800, 1800 and 2600 MHz in Asia (bands 1, 3, 5, 7, 8, 11, 13, 40) and 1800 MHz and 2300 MHz in Australia and New Zealand (bands 3, 40). As a result, phones from one country may not work in other countries. Users will need a multi-band capable phone for roaming internationally.

Patents

According to the European Telecommunications Standards Institute's (ETSI) intellectual property rights (IPR) database, about 50 companies have declared, as of March 2012, holding essential patents covering the LTE standard. The ETSI has made no investigation on the correctness of the declarations however, so that "any analysis of essential LTE patents should take into account more than ETSI declarations."

The table below shows the available LTE royalty:

Wireless Broadband

Three fixed wireless dishes with (protective covers) on top of 307 W. 7th Street, Fort Worth, Texas around 2001

Wireless broadband is technology that provides high-speed wireless Internet access or computer networking access over a wide area.

The Term Broadband

Originally the word "broadband" had a technical meaning, but became a marketing term for any kind of relatively high-speed computer network or Internet access technology. According to the 802.16-2004 standard, broadband means "having instantaneous bandwidths greater than 1 MHz and supporting data rates greater than about 1.5 Mbit/s." The Federal Communications Commission (FCC) recently re-defined the definition to mean download speeds of at least 25 Mbit/s and upload speeds of at least 3 Mbit/s.

Technology and Speeds

A typical WISP Customer Premises Equipment (CPE) installed on a residence

Wireless networks can feature data rates roughly equivalent to some wired networks, such as that of asymmetric digital subscriber line (ADSL) or a cable modem. Wireless networks can also be symmetrical, meaning the same rate in both directions (downstream and upstream), which is most commonly associated with fixed wireless networks. A fixed wireless network link is a stationary terrestrial wireless connection, which can support higher data rates for the same power as mobile or satellite systems.

Few wireless Internet service providers (WISPs) provide download speeds of over 100 Mbit/s; most broadband wireless access (BWA) services are estimated to have a range of 50 km (31 mi) from a tower. Technologies used include LMDS and MMDS, as well as heavy use of the ISM bands and one particular access technology was standardized by IEEE 802.16, with products known as WiMAX.

WiMAX is highly popular in Europe but has not met full acceptance in the United States because cost of deployment does not meet return on investment figures. In 2005 the Federal Communications Commission adopted a Report and Order that revised the FCC's rules to open the 3650 MHz band for terrestrial wireless broadband operations.

Development of Wireless Broadband in the United States

On November 14, 2007 the Commission released Public Notice DA 07-4605 in which the Wireless Telecommunications Bureau announced the start date for licensing and registration process for the 3650–3700 MHz band. In 2010 the FCC adopted the TV White Space Rules (TVWS) and allowed some of the better no line of sight frequency (700 MHz) into the FCC Part-15 Rules. The Wireless Internet Service Providers Association, a national association of WISPs, petitioned the FCC and won.

Initially, WISPs were only found in rural areas not covered by cable or DSL. These early WISPs would employ a high-capacity T-carrier, such as a T1 or DS3 connection, and then broadcast the signal from a high elevation, such as at the top of a water tower. To receive this type of Internet connection, consumers mount a small dish to the roof of their home or office and point it to the transmitter. Line of sight is usually necessary for WISPs operating in the 2.4 and 5 GHz bands with 900 MHz offering better NLOS (non-line-of-sight) performance.

Residential Wireless Internet

Providers of fixed wireless broadband services typically provide equipment to customers and install a small antenna or dish somewhere on the roof. This equipment is usually deployed as a service and maintained by the company providing that service. Fixed wireless services have become particularly popular in many rural areas where Cable, DSL or other typical home Internet services are not available.

Business Wireless Internet

Many companies in the US and worldwide have started using wireless alternatives to incumbent and local providers for internet and voice service. These providers tend to offer competitive services and options in areas where there is a difficulty getting affordable Ethernet connections from terrestrial providers such as ATT, Comcast, Verizon and others. Also, companies looking for full diversity between carriers for critical uptime requirements may seek wireless alternatives to local options.

Demand for Spectrum

To cope with increased demand for wireless broadband, increased spectrum would be needed. Studies began in 2009, and while some unused spectrum was available, it appeared broadcasters would have to give up at least some spectrum. This led to strong objections from the broadcasting community. In 2013, auctions were planned, and for now any action by broadcasters is voluntary.

Mobile Wireless Broadband

Called mobile broadband, wireless broadband technologies include services from mobile phone service providers such as Verizon Wireless, Sprint Corporation, and AT&T Mobility,and T-Mobile which allow a more mobile version of Internet access. Consumers can purchase a PC card, laptop card, or USB equipment to connect their PC or laptop to the Internet via cell phone towers. This type of connection would be stable in almost any area that could also receive a strong cell phone connection. These connections can cost more for portable convenience as well as having speed limitations in all but urban environments.

On June 2, 2010, after months of discussion, AT&T became the first wireless Internet provider in the USA to announce plans to charge according to usage. As the only iPhone service in the United States, AT&T experienced the problem of heavy Internet use more than other providers. About 3 percent of AT&T smart phone customers account for 40 percent of the technology's use. 98 percent of the company's customers use less than 2 gigabytes (4000 page views, 10,000 emails or 200 minutes of streaming video), the limit under the $25 monthly plan, and 65 percent use less than 200 megabytes, the limit for the $15 plan. For each gigabyte in excess of the limit, customers would be charged $10 a month starting June 7, 2010, though existing customers would not be required to change from the $30 a month unlimited service plan. The new plan would become a requirement for those upgrading to the new iPhone technology later in the summer.

Licensing

A wireless connection can be either licensed or unlicensed. In the US, licensed connections use a private spectrum the user has secured rights to from the Federal Communications Commission (FCC). In other countries, spectrum is licensed from the country's national radio communications authority (such as the ACMA in Australia or Nigerian Communications Commission in Nigeria (NCC)). Licensing is usually expensive and often reserved for large companies who wish to guarantee private access to spectrum for use in point to point communication. Because of this, most wireless ISP's use unlicensed spectrum which is publicly shared.

Local Multipoint Distribution Service

Local Multipoint Distribution Service (LMDS) is a broadband wireless access technology originally designed for digital television transmission (DTV). It was conceived as a fixed wireless, point-to-multipoint technology for utilization in the last mile. LMDS commonly operates on microwave frequencies across the 26 GHz and 29 GHz bands. In the United States, frequencies from 31.0 through 31.3 GHz are also considered LMDS frequencies.

Throughput capacity and reliable distance of the link depends on common radio link constraints and the modulation method used - either phase-shift keying or amplitude modulation. Distance is typically limited to about 1.5 miles (2.4 km) due to rain fade attenuation constraints. Deployment links of up to 5 miles (8 km) from the base station are possible in some circumstances such as in point-to-point systems that can reach slightly farther distances due to increased antenna gain.

History and Outlook

United States

LMDS showed great promise in the late 1990s and became known as "wireless cable" for its potential to compete with cable companies for provision of broadband television to the home. The Federal Communications Commission auctioned spectrum for LMDS in 1998 and 1999.

Despite its early potential and the hype that surrounded the technology, LMDS was slow to find commercial traction. Many equipment and technology vendors simply abandoned their LMDS product portfolios.

Industry observers believe that the window for LMDS has closed with newer technologies replacing it. Major telecommunications companies have been aggressive about deploying alternative technologies such as IPTV and fiber to the premises, also called "fiber optics". Moreover, LMDS has been surpassed in both technological and commercial potential by LTE and WiMax standards.

Europe and Worldwide

Although some operators use LMDS to provide access services, LMDS is more commonly used for high-capacity backhaul for interconnection of networks such as GSM, UMTS, WiMAX and Wi-Fi.

References

- Hsiao-Hwa Chen (2007), The Next Generation CDMA Technologies, John Wiley and Sons, pp. 105–106, ISBN 978-0-470-02294-8

- Phil Goldstein (22 June 2012). "Report: TD-LTE to power 25% of LTE connections by 2016". FierceWireless. Retrieved 10 December 2013.

- "Xinwei finally stages user trials; will trade under CooTel brand". TeleGeography. 2016-01-19. Retrieved 2016-01-20.

- "T-Mobile shifting 1700 MHz HSPA+ users to 1900 MHz band". TeleGeography. 2015-06-24. Retrieved 2016-04-07.

- Sam Byford (20 February 2012). "SoftBank launching 110Mbps AXGP 4G network in Japan this week". The Verge. Retrieved 7 June 2015.

- Zahid Ghadialy (21 February 2012). "SoftBank launching 110Mbps AXGP 4G network in Japan this week". The 3G4G Blog. Retrieved 7 June 2015.

- "LTE Direct: A thriving ecosystem and amazing use cases on full display at MWC 2015 [VIDEOS] | Qualcomm". Qualcomm. Retrieved 2015-10-20.

- Siemens (2004-06-10). "TD-SCDMA Whitepaper: the Solution for TDD bands" (PDF). TD Forum. pp. 6–9. Archived from the original (pdf) on 2014-03-30. Retrieved 2009-06-15.

- "China Mobile Announces Commercial Deployment of TD-SCDMA Technology". Spreadtrum Communications, Inc. 2008-03-28. Retrieved 2014-07-17.

- Craig Timberg (24 August 2014). "For sale: Systems that can secretly track where cellphone users go around the globe". Washington Post. Retrieved 20 December 2014.

- Craig Timberg (18 December 2014). "German researchers discover a flaw that could let anyone listen to your cell calls.". The Switch- Washington Post. Washington Post. Retrieved 20 December 2014.

- Peter Onneken (18 December 2014). "Sicherheitslücken im UMTS-Netz". Tagesschau (in German). ARD-aktuell / tagesschau.de. Retrieved 20 December 2014.

- Josh Taylor (4 December 2012). "Optus to launch TD-LTE 4G network in Canberra". ZDNet. Retrieved 9 January 2014.

- "MWC 2013: Ericsson and China Mobile demo first dual mode HD VoLTE call based on multi-mode chipsets". Wireless - Wireless Communications For Public Services And Private Enterprises. Noble House Media. 4 March 2013. Retrieved 9 January 2014.

- Steve Costello (2 August 2013). "GCF and GTI partner for TD-LTE device certification". Mobile World Live. Retrieved 9 January 2014.

- Ben Munson (31 January 2014). "China Mobile, NSN Complete Live VoLTE Test on TD-LTE". Wireless Week. Retrieved 11 February 2014.

- "NSN and Sprint achieves huge leap in TD-LTE network speeds". TelecomTiger. 6 February 2014. Retrieved 11 February 2014.

- Dan Jones (16 October 2012). "Defining 4G: What the Heck Is LTE TDD?". Light Reading. Retrieved 9 January 2014.

- "Huawei partners with Aero2 to launch LTE TDD/FDD commercial network". Computer News Middle East. 21 September 2011. Retrieved 10 December 2013.

- Kevin Fitchard (30 October 2013). "What's igniting Spark? A look inside Sprint's super-LTE network". GigaOM. Retrieved 10 December 2013.

Voice Over IP and Various Protocols

VoIP technology provides voice communications over the Internet and other Internet Protocol networks. These technologies enhance WiMAX technologies. This chapter is dedicated to the examination of Media Gateway Control Protocol, Simple Gateway Control Protocol, IEEE 802.21, IEEE 802.11u, Session Initiation Protocol, Mobile VoIP etc. This section provides the reader with a comprehensive study of technologies using VoIP.

Voice Over IP

Voice over Internet Protocol (Voice over IP, VoIP and IP telephony) is a methodology and group of technologies for the delivery of voice communications and multimedia sessions over Internet Protocol (IP) networks, such as the Internet. The terms Internet telephony, broadband telephony, and broadband phone service specifically refer to the provisioning of communications services (voice, fax, SMS, voice-messaging) over the public Internet, rather than via the public switched telephone network (PSTN).

The steps and principles involved in originating VoIP telephone calls are similar to traditional digital telephony and involve signaling, channel setup, digitization of the analog voice signals, and encoding. Instead of being transmitted over a circuit-switched network; however, the digital information is packetized, and transmission occurs as IP packets over a packet-switched network. They transport audio streams using special media delivery protocols that encode audio and video with audio codecs, and video codecs. Various codecs exist that optimize the media stream based on application requirements and network bandwidth; some implementations rely on narrowband and compressed speech, while others support high fidelity stereo codecs. Some popular codecs include μ-law and a-law versions of G.711, G.722, a popular open source voice codec known as iLBC, a codec that only uses 8 kbit/s each way called G.729, and many others.

Early providers of voice-over-IP services offered business models and technical solutions that mirrored the architecture of the legacy telephone network. Second-generation providers, such as Skype, have built closed networks for private user bases, offering the benefit of free calls and convenience while potentially charging for access to other communication networks, such as the PSTN. This has limited the freedom of users to mix-and-match third-party hardware and software. Third-generation providers, such as Google Talk, have adopted the concept of federated VoIP—which is a departure from the architecture of the legacy networks. These solutions typically allow dynamic interconnection between users on any two domains on the Internet when a user wishes to place a call.

In addition to VoIP phones, VoIP is available on many smartphones, personal computers, and on Internet access devices. Calls and SMS text messages may be sent over 3G/4G or Wi-Fi.

Pronunciation

VoIP is variously pronounced as initials, *V-O-I-P*, or as an acronym, usually (*voyp*), as in *voice*, but pronunciation in full words, *voice over Internet Protocol*, and *voice over IP*, are common.

Protocols

Voice over IP has been implemented in various ways using both proprietary protocols and protocols based on open standards. VoIP protocols include:

- Session Initiation Protocol (SIP)
- H.323
- Media Gateway Control Protocol (MGCP)
- Gateway Control Protocol (Megaco, H.248)
- Real-time Transport Protocol (RTP)
- Real-time Transport Control Protocol (RTCP)
- Secure Real-time Transport Protocol (SRTP)
- Session Description Protocol (SDP)
- Inter-Asterisk eXchange (IAX)
- Jingle XMPP VoIP extensions
- Skype protocol
- Teamspeak

The H.323 protocol was one of the first VoIP protocols that found widespread implementation for long-distance traffic, as well as local area network services. However, since the development of newer, less complex protocols such as MGCP and SIP, H.323 deployments are increasingly limited to carrying existing long-haul network traffic.

These protocols can be used by special-purpose software, such as Jitsi, or integrated into a web page (web-based VoIP), like Google Talk.

Adoption

Consumer Market

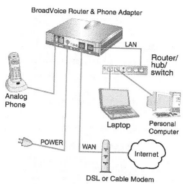

Example of residential network including VoIP

A major development that started in 2004 was the introduction of mass-market VoIP services that utilize existing broadband Internet access, by which subscribers place and receive telephone calls in much the same manner as they would via the public switched telephone network (PSTN). Full-service VoIP phone companies provide inbound and outbound service with direct inbound dialing. Many offer unlimited domestic calling for a flat monthly subscription fee. This sometimes includes international calls to certain countries. Phone calls between subscribers of the same provider are usually free when flat-fee service is not available. A VoIP phone is necessary to connect to a VoIP service provider. This can be implemented in several ways:

- Dedicated VoIP phones connect directly to the IP network using technologies such as wired Ethernet or Wi-Fi. They are typically designed in the style of traditional digital business telephones.

- An analog telephone adapter is a device that connects to the network and implements the electronics and firmware to operate a conventional analog telephone attached through a modular phone jack. Some residential Internet gateways and cablemodems have this function built in.

- A softphone is application software installed on a networked computer that is equipped with a microphone and speaker, or headset. The application typically presents a dial pad and display field to the user to operate the application by mouse clicks or keyboard input.

PSTN and Mobile Network Providers

It is becoming increasingly common for telecommunications providers to use VoIP telephony over dedicated and public IP networks to connect switching centers and to interconnect with other telephony network providers; this is often referred to as "IP backhaul".

Smartphones and Wi-Fi-enabled mobile phones may have SIP clients built into the firmware or available as an application download.

Corporate Use

Because of the bandwidth efficiency and low costs that VoIP technology can provide, businesses are migrating from traditional copper-wire telephone systems to VoIP systems to reduce their monthly phone costs. In 2008, 80% of all new Private branch exchange (PBX) lines installed internationally were VoIP.

VoIP solutions aimed at businesses have evolved into unified communications services that treat all communications—phone calls, faxes, voice mail, e-mail, Web conferences, and more—as discrete units that can all be delivered via any means and to any handset, including cellphones. Two kinds of competitors are competing in this space: one set is focused on VoIP for medium to large enterprises, while another is targeting the small-to-medium business (SMB) market.

VoIP allows both voice and data communications to be run over a single network, which can significantly reduce infrastructure costs.

The prices of extensions on VoIP are lower than for PBX and key systems. VoIP switches may run on commodity hardware, such as personal computers. Rather than closed architectures, these devices rely on standard interfaces.

VoIP devices have simple, intuitive user interfaces, so users can often make simple system configuration changes. Dual-mode phones enable users to continue their conversations as they move between an outside cellular service and an internal Wi-Fi network, so that it is no longer necessary to carry both a desktop phone and a cellphone. Maintenance becomes simpler as there are fewer devices to oversee.

Skype, which originally marketed itself as a service among friends, has begun to cater to businesses, providing free-of-charge connections between any users on the Skype network and connecting to and from ordinary PSTN telephones for a charge.

In the United States the Social Security Administration (SSA) is converting its field offices of 63,000 workers from traditional phone installations to a VoIP infrastructure carried over its existing data network.

Quality of Service

Communication on the IP network is perceived as less reliable in contrast to the circuit-switched public telephone network because it does not provide a network-based mechanism to ensure that data packets are not lost, and are delivered in sequential order. It is a best-effort network without fundamental Quality of Service (QoS) guarantees. Therefore, VoIP implementations may face problems with latency, packet loss, and jitter.

By default, network routers handle traffic on a first-come, first-served basis. Network routers on high volume traffic links may introduce latency that exceeds permissible thresholds for VoIP. Fixed delays cannot be controlled, as they are caused by the physical distance the packets travel; however, latency can be minimized by marking voice packets as being delay-sensitive with methods such as DiffServ.

VoIP endpoints usually have to wait for completion of transmission of previous packets before new data may be sent. Although it is possible to preempt (abort) a less important packet in mid-transmission, this is not commonly done, especially on high-speed links where transmission times are short even for maximum-sized packets. An alternative to preemption on slower links, such as dialup and digital subscriber line (DSL), is to reduce the maximum transmission time by reducing the maximum transmission unit. But every packet must contain protocol headers, so this increases relative header overhead on every link traversed, not just the bottleneck (usually Internet access) link.

DSL modems provide Ethernet (or Ethernet over USB) connections to local equipment, but inside they are actually Asynchronous Transfer Mode (ATM) modems. They use ATM Adaptation Layer 5 (AAL5) to segment each Ethernet packet into a series of 53-byte ATM cells for transmission, reassembling them back into Ethernet frames at the receiving end. A virtual circuit identifier (VCI) is part of the 5-byte header on every ATM cell, so the transmitter can multiplex the active virtual circuits (VCs) in any arbitrary order. Cells from the *same* VC are always sent sequentially.

However, a majority of DSL providers use only one VC for each customer, even those with bundled VoIP service. Every Ethernet frame must be completely transmitted before another can begin. If a second VC were established, given high priority and reserved for VoIP, then a low priority data packet could be suspended in mid-transmission and a VoIP packet sent right away on the high priority VC. Then the link would pick up the low priority VC where it left off. Because ATM links are multiplexed on a cell-by-cell basis, a high priority packet would have to wait at most 53 byte times to begin transmission. There would be no need to reduce the interface MTU and accept the resulting increase in higher layer protocol overhead, and no need to abort a low priority packet and resend it later.

ATM has substantial header overhead: $5/53 = 9.4\%$, roughly twice the total header overhead of a 1500 byte Ethernet frame. This "ATM tax" is incurred by every DSL user whether or not they take advantage of multiple virtual circuits - and few can.

ATM's potential for latency reduction is greatest on slow links, because worst-case latency decreases with increasing link speed. A full-size (1500 byte) Ethernet frame takes 94 ms to transmit at 128 kbit/s but only 8 ms at 1.5 Mbit/s. If this is the bottleneck link, this latency is probably small enough to ensure good VoIP performance without MTU reductions or multiple ATM VCs. The latest generations of DSL, VDSL and VDSL2, carry Ethernet without intermediate ATM/AAL5 layers, and they generally support IEEE 802.1p priority tagging so that VoIP can be queued ahead of less time-critical traffic.

Voice, and all other data, travels in packets over IP networks with fixed maximum capacity. This system may be more prone to congestion and DoS attacks than traditional circuit switched systems; a circuit switched system of insufficient capacity will refuse new connections while carrying the remainder without impairment, while the quality of real-time data such as telephone conversations on packet-switched networks degrades dramatically.

Fixed delays cannot be controlled as they are caused by the physical distance the packets travel. They are especially problematic when satellite circuits are involved because of the long distance to a geostationary satellite and back; delays of 400–600 ms are typical.

When the load on a link grows so quickly that its switches experience queue overflows, congestion results and data packets are lost. This signals a transport protocol like TCP to reduce its transmission rate to alleviate the congestion. But VoIP usually uses UDP not TCP because recovering from congestion through retransmission usually entails too much latency. So QoS mechanisms can avoid the undesirable loss of VoIP packets by immediately transmitting them ahead of any queued bulk traffic on the same link, even when that bulk traffic queue is overflowing.

The receiver must resequence IP packets that arrive out of order and recover gracefully when packets arrive too late or not at all. Jitter results from the rapid and random (i.e. unpredictable) changes in queue lengths along a given Internet path due to competition from other users for the same transmission links. VoIP receivers counter jitter by storing incoming packets briefly in a "de-jitter" or "playout" buffer, deliberately increasing latency to improve the chance that each packet will be on hand when it is time for the voice engine to play it. The added delay is thus a compromise between excessive latency and excessive dropout, i.e. momentary audio interruptions.

Although jitter is a random variable, it is the sum of several other random variables that are at least somewhat independent: the individual queuing delays of the routers along the Internet path in question. Thus according to the central limit theorem, we can model jitter as a gaussian random variable. This suggests continually estimating the mean delay and its standard deviation and setting the playout delay so that only packets delayed more than several standard deviations above the mean will arrive too late to be useful. In practice, however, the variance in latency of many Internet paths is dominated by a small number (often one) of relatively slow and congested "bottleneck" links. Most Internet backbone links are now so fast (e.g. 10 Gbit/s) that their delays are dominated by the transmission medium (e.g. optical fiber) and the routers driving them do not have enough buffering for queuing delays to be significant.

It has been suggested to rely on the packetized nature of media in VoIP communications and transmit the stream of packets from the source phone to the destination phone simultaneously across different routes (multi-path routing). In such a way, temporary failures have less impact on the communication quality. In capillary routing it has been suggested to use at the packet level Fountain codes or particularly raptor codes for transmitting extra redundant packets making the communication more reliable.

A number of protocols have been defined to support the reporting of quality of service (QoS) and quality of experience (QoE) for VoIP calls. These include RTCP Extended Report (RFC 3611), SIP RTCP Summary Reports, H.460.9 Annex B (for H.323), H.248.30 and MGCP extensions. The RFC 3611 VoIP Metrics block is generated by an IP phone or gateway during a live call and contains information on packet loss rate, packet discard rate (because of jitter), packet loss/discard burst metrics (burst length/density, gap length/density), network delay, end system delay, signal / noise / echo level, Mean Opinion Scores (MOS) and R factors and configuration information related to the jitter buffer.

RFC 3611 VoIP metrics reports are exchanged between IP endpoints on an occasional basis during a call, and an end of call message sent via SIP RTCP Summary Report or one of the other signaling protocol extensions. RFC 3611 VoIP metrics reports are intended to support real time feedback related to QoS problems, the exchange of information between the endpoints for improved call quality calculation and a variety of other applications.

Rural areas in particular are greatly hindered in their ability to choose a VoIP system over PBX. This is generally down to the poor access to superfast broadband in rural country areas. With the release of 4G data, there is a potential for corporate users based outside of populated areas to switch their internet connection to 4G data, which is comparatively as fast as a regular superfast broadband connection. This greatly enhances the overall quality and user experience of a VoIP system in these areas. This method was already trialled in rural Germany, surpassing all expectations.

Layer 2

A number of protocols that deal with the data link layer and physical layer include quality-of-service mechanisms that can be used to ensure that applications like VoIP work well even in congested scenarios. Some examples include:

- IEEE 802.11e is an approved amendment to the IEEE 802.11 standard that defines a set of quality-of-service enhancements for wireless LAN applications through modifications to the Media Access Control (MAC) layer. The standard is considered of critical importance for delay-sensitive applications, such as voice over wireless IP.

- IEEE 802.1p defines 8 different classes of service (including one dedicated to voice) for traffic on layer-2 wired Ethernet.

- The ITU-T G.hn standard, which provides a way to create a high-speed (up to

1 gigabit per second) Local area network (LAN) using existing home wiring (power lines, phone lines and coaxial cables). G.hn provides QoS by means of "Contention-Free Transmission Opportunities" (CFTXOPs) which are allocated to flows (such as a VoIP call) which require QoS and which have negotiated a "contract" with the network controllers.

VoIP Performance Metrics

The quality of voice transmission is characterized by several metrics that may be monitored by network elements, by the user agent hardware or software. Such metrics include network packet loss, packet jitter, packet latency (delay), post-dial delay, and echo. The metrics are determined by VoIP performance testing and monitoring.

PSTN Integration

The Media VoIP Gateway connects the digital media stream, so as to complete creating the path for voice as well as data media. It includes the interface for connecting the standard PSTN networks with the ATM and Inter Protocol networks. The Ethernet interfaces are also included in the modern systems, which are specially designed to link calls that are passed via the VoIP.

E.164 is a global FGFnumbering standard for both the PSTN and PLMN. Most VoIP implementations support E.164 to allow calls to be routed to and from VoIP subscribers and the PSTN/PLMN. VoIP implementations can also allow other identification techniques to be used. For example, Skype allows subscribers to choose "Skype names" (usernames) whereas SIP implementations can use URIs similar to email addresses. Often VoIP implementations employ methods of translating non-E.164 identifiers to E.164 numbers and vice versa, such as the Skype-In service provided by Skype and the ENUM service in IMS and SIP.

Echo can also be an issue for PSTN integration. Common causes of echo include impedance mismatches in analog circuitry and acoustic coupling of the transmit and receive signal at the receiving end.

Number Portability

Local number portability (LNP) and Mobile number portability (MNP) also impact VoIP business. In November 2007, the Federal Communications Commission in the United States released an order extending number portability obligations to interconnected VoIP providers and carriers that support VoIP providers. Number portability is a service that allows a subscriber to select a new telephone carrier without requiring a new number to be issued. Typically, it is the responsibility of the former carrier to "map" the old number to the undisclosed number assigned by the new carrier. This is achieved by maintaining a database of numbers. A dialed number is initially received by the original carrier and quickly rerouted to the new carrier. Multiple porting references must be maintained even if the subscriber returns to the original carrier. The FCC mandates carrier compliance with these consumer-protection stipulations.

A voice call originating in the VoIP environment also faces challenges to reach its destination if the number is routed to a mobile phone number on a traditional mobile carrier. VoIP has been identified in the past as a Least Cost Routing (LCR) system, which is based on checking the destination of each telephone call as it is made, and then sending the call via the network that will cost the customer the least. This rating is subject to some debate given the complexity of call routing created by number portability. With GSM number portability now in place, LCR providers can no longer rely on using the network root prefix to determine how to route a call. Instead, they must now determine the actual network of every number before routing the call.

Therefore, VoIP solutions also need to handle MNP when routing a voice call. In countries without a central database, like the UK, it might be necessary to query the GSM network about which home network a mobile phone number belongs to. As the popularity of VoIP increases in the enterprise markets because of least cost routing options, it needs to provide a certain level of reliability when handling calls.

MNP checks are important to assure that this quality of service is met. Handling MNP lookups before routing a call provides some assurance that the voice call will actually work.

Emergency Calls

A telephone connected to a land line has a direct relationship between a telephone number and a physical location, which is maintained by the telephone company and available to emergency responders via the national emergency response service centers in form of emergency subscriber lists. When an emergency call is received by a center the location is automatically determined from its databases and displayed on the operator console.

In IP telephony, no such direct link between location and communications end point exists. Even a provider having hardware infrastructure, such as a DSL provider, may only know the approximate location of the device, based on the IP address allocated to the network router and the known service address. However, some ISPs do not track the automatic assignment of IP addresses to customer equipment.

IP communication provides for device mobility. For example, a residential broadband connection may be used as a link to a virtual private network of a corporate entity, in which case the IP address being used for customer communications may belong to the enterprise, not being the network address of the residential ISP. Such off-premises extensions may appear as part of an upstream IP PBX. On mobile devices, e.g., a 3G handset or USB wireless broadband adapter, the IP address has no relationship with any physical location known to the telephony service provider, since a mobile user could be anywhere in a region with network coverage, even roaming via another cellular company.

At the VoIP level, a phone or gateway may identify itself with a Session Initiation Protocol (SIP) registrar by its account credentials. In such cases, the Internet telephony service provider (ITSP) only knows that a particular user's equipment is active. Service providers often provide emergency response services by agreement with the user who registers a physical location and agrees that emergency services are only provided to that address if an emergency number is called from the IP device.

Such emergency services are provided by VoIP vendors in the United States by a system called Enhanced 911 (E911), based on the Wireless Communications and Public Safety Act of 1999. The VoIP E911 emergency-calling system associates a physical address with the calling party's telephone number. All VoIP providers that provide access to the public switched telephone network are required to implement E911, a service for which the subscriber may be charged. However, end-customer participation in E911 is not mandatory and customers may opt out of the service.

The VoIP E911 system is based on a static table lookup. Unlike in cellular phones, where the location of an E911 call can be traced using assisted GPS or other methods, the VoIP E911 information is only accurate so long as subscribers, who have the legal responsibility, are diligent in keeping their emergency address information current.

Fax Support

Support for fax has been problematic in many VoIP implementations, as most voice digitization and compression codecs are optimized for the representation of the human voice and the proper timing of the modem signals cannot be guaranteed in a packet-based, connection-less network. An alternative IP-based solution for delivering fax-over-IP called T.38 is available. Sending faxes using VoIP is sometimes referred to as FoIP, or Fax over IP.

The T.38 protocol is designed to compensate for the differences between traditional packet-less communications over analog lines and packet-based transmissions which are the basis for IP communications. The fax machine could be a traditional fax machine connected to the PSTN, or an ATA box (or similar). It could be a fax machine with an RJ-45 connector plugged straight into an IP network, or it could be a computer pretending to be a fax machine. Originally, T.38 was designed to use UDP and TCP transmission methods across an IP network. TCP is better suited for use between two IP devices. However, older fax machines, connected to an analog system, benefit from UDP near real-time characteristics due to the "no recovery rule" when a UDP packet is lost or an error occurs during transmission. UDP transmissions are preferred as they do not require testing for dropped packets and as such since each T.38 packet transmission includes a majority of the data sent in the prior packet, a T.38 termination point has a higher degree of success in re-assembling the fax transmission back into its original form for interpretation by the end device. This in an attempt to overcome the obstacles of simulating real time transmissions using packet based protocol.

There have been updated versions of T.30 to resolve the fax over IP issues, which is the core fax protocol. Some newer high end fax machines have T.38 built-in capabilities which allow the user to plug right into the network and transmit/receive faxes in native T.38 like the Ricoh 4410NF Fax Machine. A unique feature of T.38 is that each packet contains a portion of the main data sent in the previous packet. With T.38, two successive lost packets are needed to actually lose any data. The data one will lose will only be a small piece, but with the right settings and error correction mode, there is an increased likelihood that they will receive enough of the transmission to satisfy the requirements of the fax machine for output of the sent document.

While many late-model analog telephone adapters (ATAs) support T.38, uptake has been limited as many voice-over-IP providers perform least-cost routing which selects the least expensive PSTN gateway in the called city for an outbound message. There is typically no means to ensure that that gateway is T.38 capable. Providers often place their own equipment (such as an Asterisk PBX installation) in the signal path, which creates additional issues as every link in the chain must be T.38 aware for the protocol to work. Similar issues arise if a provider is purchasing local direct inward dial numbers from the lowest bidder in each city, as many of these may not be T.38 enabled.

Power Requirements

Telephones for traditional residential analog service are usually connected directly to telephone company phone lines which provide direct current to power most basic analog handsets independently of locally available electrical power.

IP Phones and VoIP telephone adapters connect to routers or cable modems which typically depend on the availability of mains electricity or locally generated power. Some VoIP service providers use customer premises equipment (e.g., cablemodems) with battery-backed power supplies to assure uninterrupted service for up to several hours in case of local power failures. Such battery-backed devices typically are designed for use with analog handsets.

Some VoIP service providers implement services to route calls to other telephone services of the subscriber, such a cellular phone, in the event that the customer's network device is inaccessible to terminate the call.

The susceptibility of phone service to power failures is a common problem even with traditional analog service in areas where many customers purchase modern telephone units that operate with wireless handsets to a base station, or that have other modern phone features, such as built-in voicemail or phone book features.

Security

The security concerns of VoIP telephone systems are similar to those of any Internet-connected device. This means that hackers who know about these vulnerabilities can institute denial-of-service attacks, harvest customer data, record conversations and compromise voicemail messages. The quality of internet connection determines the quality of the calls. VoIP phone service also will not work if there is power outage and when the internet connection is down. The 9-1-1 or 112 service provided by VoIP phone service is also different from analog phone which is associated with a fixed address. The emergency center may not be able to determine your location based on your virtual phone number. Compromised VoIP user account or session credentials may enable an attacker to incur substantial charges from third-party services, such as long-distance or international telephone calling.

The technical details of many VoIP protocols create challenges in routing VoIP traffic through firewalls and network address translators, used to interconnect to transit networks or the Internet. Private session border controllers are often employed to enable VoIP calls to and from protected networks. Other methods to traverse NAT devices involve assistive protocols such as STUN and Interactive Connectivity Establishment (ICE).

Many consumer VoIP solutions do not support encryption of the signaling path or the media, however securing a VoIP phone is conceptually easier to implement than on traditional telephone circuits. A result of the lack of encryption is a relative easy to eavesdrop on VoIP calls when access to the data network is possible. Free open-source solutions, such as Wireshark, facilitate capturing VoIP conversations.

Standards for securing VoIP are available in the Secure Real-time Transport Protocol (SRTP) and the ZRTP protocol for analog telephony adapters as well as for some softphones. IPsec is available to secure point-to-point VoIP at the transport level by using opportunistic encryption.

Government and military organizations use various security measures to protect VoIP traffic, such as voice over secure IP (VoSIP), secure voice over IP (SVoIP), and secure voice over secure IP (SVoSIP). The distinction lies in whether encryption is applied in the telephone or in the network or both. Secure voice over secure IP is accomplished by encrypting VoIP with protocols such as SRTP or ZRTP. Secure voice over IP is accomplished by using Type 1 encryption on a classified network, like SIPRNet. Public Secure VoIP is also available with free GNU programs and in many popular commercial VoIP programs via libraries such as ZRTP.

Caller ID

Caller ID support among VoIP providers varies, but is provided by the majority of VoIP providers. Many VoIP service providers allow callers to configure arbitrary caller ID information, thus permitting spoofing attacks. Business-grade VoIP equipment and software often makes it easy to modify caller ID information, providing many businesses great flexibility.

The United States enacted the Truth in Caller ID Act of 2009 on December 22, 2010. This law makes it a crime to "knowingly transmit misleading or inaccurate caller identification information with the intent to defraud, cause harm, or wrongfully obtain anything of value ...". Rules implementing the law were adopted by the Federal Communications Commission on June 20, 2011.

Compatibility with Traditional Analog Telephone Sets

Most analog telephone adapters do not decode dial pulses generated by older telephones, supporting only touch-tone. Pulse-to-tone converters are commercially available; a user reports that a few specific ATA models (such as the Grandstream 502) recognise pulse dial directly, but are poorly documented and provide no assurance that newer models in the same series will retain this compatibility. This however, will only work for one VoIP conversation at a time.

Support for Other Telephony Devices

Another challenge for VoIP implementations is the proper handling of outgoing calls from other telephony devices such as digital video recorders, satellite television receivers, alarm systems, conventional modems and other similar devices that depend on access to a PSTN telephone line for some or all of their functionality.

These types of calls sometimes complete without any problems, but in other cases they fail. If VoIP and cellular substitution becomes very popular, some ancillary equipment makers may be forced to redesign equipment, because it would no longer be possible to assume a conventional PSTN telephone line would be available in consumers' houses.

Operational Cost

VoIP can be a benefit for reducing communication and infrastructure costs. Examples include:

- Routing phone calls over existing data networks to avoid the need for separate voice and data networks.

- The ability to transmit more than one telephone call over a single broadband connection.

- Secure calls using standardized protocols (such as Secure Real-time Transport Protocol). Most of the difficulties of creating a secure telephone connection over traditional phone lines, such as digitizing and digital transmission, are already in place with VoIP. It is only necessary to encrypt and authenticate the existing data stream.

- Utilized existing network infrastructure to minimize the operating cost.

- Eliminating the need of hiring personnel to greet and distribute incoming calls with the use of a Virtual PBX

Regulatory and Legal Issues

As the popularity of VoIP grows, governments are becoming more interested in regulating VoIP in a manner similar to PSTN services.

Throughout the developing world, countries where regulation is weak or captured by the dominant operator, restrictions on the use of VoIP are imposed, including in Panama where VoIP is taxed, Guyana where VoIP is prohibited and India where its retail commercial sales is allowed but only for long distance service. In Ethiopia, where the government is nationalising telecommunication service, it is a criminal offence to offer services using VoIP. The country has installed firewalls to prevent international calls being made using VoIP. These measures were taken after the popularity of VoIP reduced the income generated by the state owned telecommunication company.

European Union

In the European Union, the treatment of VoIP service providers is a decision for each national telecommunications regulator, which must use competition law to define relevant national markets and then determine whether any service provider on those national markets has "significant market power" (and so should be subject to certain obligations). A general distinction is usually made between VoIP services that function over managed networks (via broadband connections) and VoIP services that function over unmanaged networks (essentially, the Internet).

The relevant EU Directive is not clearly drafted concerning obligations which can exist independently of market power (e.g., the obligation to offer access to emergency calls), and it is impossible to say definitively whether VoIP service providers of either type are bound by them. A review of the EU Directive is under way and should be complete by 2007.

Middle East

In the UAE and Oman it is illegal to use any form of VoIP, to the extent that Web sites of Gizmo5 are blocked. Providing or using VoIP services is illegal in Oman. Those who violate the law stand to be fined 50,000 Omani Rial (about 130,317 US dollars) or spend two years in jail or both. In 2009, police in Oman have raided 121 Internet cafes throughout the country and arrested 212 people for using/providing VoIP services.

India

In India, it is legal to use VoIP, but it is illegal to have VoIP gateways inside India. This effectively means that people who have PCs can use them to make a VoIP call to any number, but if the remote side is a normal phone, the gateway that converts the VoIP call to a POTS call is not permitted by law to be inside India. Foreign based Voip server services are illegal to use in India.

In the interest of the Access Service Providers and International Long Distance Operators the Internet telephony was permitted to the ISP with restrictions. Internet Telephony is considered to be different service in its scope, nature and kind from real time voice as offered by other Access Service Providers and Long Distance Carriers. Hence the following type of Internet Telephony are permitted in India:

(a) PC to PC; within or outside India

(b) PC / a device / Adapter conforming to standard of any international agencies like- ITU or IETF etc. in India to PSTN/PLMN abroad.

(c) Any device / Adapter conforming to standards of International agencies like ITU, IETF etc. connected to ISP node with static IP address to similar device / Adapter; within or outside India.

(d) Except whatever is described in condition (ii) above, no other form of Internet Telephony is permitted.

(e) In India no Separate Numbering Scheme is provided to the Internet Telephony. Presently the 10 digit Numbering allocation based on E.164 is permitted to the Fixed Telephony, GSM, CDMA wireless service. For Internet Telephony the numbering scheme shall only conform to IP addressing Scheme of Internet Assigned Numbers Authority (IANA). Translation of E.164 number / private number to IP address allotted to any device and vice versa, by ISP to show compliance with IANA numbering scheme is not permitted.

(f) The Internet Service Licensee is not permitted to have PSTN/PLMN connectivity. Voice communication to and from a telephone connected to PSTN/PLMN and following E.164 numbering is prohibited in India.

South Korea

In South Korea, only providers registered with the government are authorized to offer VoIP services. Unlike many VoIP providers, most of whom offer flat rates, Korean VoIP services are generally metered and charged at rates similar to terrestrial calling. Foreign VoIP providers encounter high barriers to government registration. This issue came to a head in 2006 when Internet service providers providing personal Internet services by contract to United States Forces Korea members residing on USFK bases threatened to block off access to VoIP services used by USFK members as an economical way to keep in contact with their families in the United States, on the grounds that the service members' VoIP providers were not registered. A compromise was reached between USFK and Korean telecommunications officials in January 2007, wherein USFK service members arriving in Korea before June 1, 2007, and subscribing to the ISP services provided on base may continue to use their US-based VoIP subscription, but later arrivals must use a Korean-based VoIP provider, which by contract will offer pricing similar to the flat rates offered by US VoIP providers.

United States

In the United States, the Federal Communications Commission requires all interconnected VoIP service providers to comply with requirements comparable to those for traditional telecommunications service providers. VoIP operators in the US are required to support local number portability; make service accessible to people with disabilities; pay regulatory fees, universal service contributions, and other mandated payments; and enable law enforcement authorities to conduct surveillance pursuant to the Communications Assistance for Law Enforcement Act (CALEA).

Operators of "Interconnected" VoIP (fully connected to the PSTN) are mandated to provide Enhanced 911 service without special request, provide for customer location updates, clearly disclose any limitations on their E-911 functionality to their consumers, obtain affirmative acknowledgements of these disclosures from all consumers, and 'may not allow their customers to "opt-out" of 911 service.' VoIP operators also receive the benefit of certain US telecommunications regulations, including an entitlement to interconnection and exchange of traffic with incumbent local exchange carriers via wholesale carriers. Providers of "nomadic" VoIP service—those who are unable to determine the location of their users—are exempt from state telecommunications regulation.

Another legal issue that the US Congress is debating concerns changes to the Foreign Intelligence Surveillance Act. The issue in question is calls between Americans and foreigners. The National Security Agency (NSA) is not authorized to tap Americans' conversations without a warrant—but the Internet, and specifically VoIP does not draw as clear a line to the location of a caller or a call's recipient as the traditional phone system does. As VoIP's low cost and flexibility convinces more and more organizations to adopt the technology, the surveillance for law enforcement agencies becomes more difficult. VoIP technology has also increased security concerns because VoIP and similar technologies have made it more difficult for the government to determine where a target is physically located when communications are being intercepted, and that creates a whole set of new legal challenges.

Internet Protocol

The Internet Protocol (IP) is the principal communications protocol in the Internet protocol suite for relaying datagrams across network boundaries. Its routing function enables internetworking, and essentially establishes the Internet.

IP has the task of delivering packets from the source host to the destination host solely based on the IP addresses in the packet headers. For this purpose, IP defines packet structures that encapsulate the data to be delivered. It also defines addressing methods that are used to label the datagram with source and destination information.

Historically, IP was the connectionless datagram service in the original *Transmission Control Program* introduced by Vint Cerf and Bob Kahn in 1974; the other being the connection-oriented Transmission Control Protocol (TCP). The Internet protocol suite is therefore often referred to as TCP/IP.

The first major version of IP, Internet Protocol Version 4 (IPv4), is the dominant protocol of the Internet. Its successor is Internet Protocol Version 6 (IPv6).

Function

The Internet Protocol is responsible for addressing hosts and for routing datagrams (packets) from a source host to a destination host across one or more IP networks. For this purpose, the Internet Protocol defines the format of packets and provides an addressing system that has two functions: Identifying hosts and providing a logical location service.

Datagram Construction

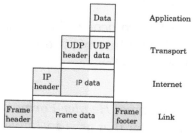

Sample encapsulation of application data from UDP to a Link protocol frame

Each datagram has two components: a header and a payload. The IP header is tagged with the source IP address, the destination IP address, and other meta-data needed to route and deliver the datagram. The payload is the data that is transported. This method of nesting the data payload in a packet with a header is called encapsulation.

IP Addressing and Routing

IP addressing entails the assignment of IP addresses and associated parameters to host interfaces. The address space is divided into networks and subnetworks, involving the designation of network or routing prefixes. IP routing is performed by all hosts, as well as routers, whose main function is to transport packets across network boundaries. Routers communicate with one another via specially designed routing protocols, either interior gateway protocols or exterior gateway protocols, as needed for the topology of the network.

IP routing is also common in local networks. For example, many Ethernet switches support IP multicast operations. These switches use IP addresses and Internet Group Management Protocol to control multicast routing but use MAC addresses for the actual routing.

Version History

The versions currently relevant are IPv4 and IPv6.

In May 1974, the Institute of Electrical and Electronic Engineers (IEEE) published a paper entitled "A Protocol for Packet Network Intercommunication". The paper's authors, Vint Cerf and Bob Kahn, described an internetworking protocol for sharing resources using packet switching among network nodes. A central control component of this model was the "Transmission Control Program" that incorporated both connection-oriented links and datagram services between hosts. The monolithic Transmission Control Program was later divided into a modular architecture consisting of the Transmission Control Protocol at the transport layer and the Internet Protocol at the network layer. The model became known as the *Department of Defense (DoD) Internet Model* and *Internet Protocol Suite*, and informally as *TCP/IP*.

IP versions 0 to 3 were experimental versions, used between 1977 and 1979. The following Internet Experiment Note (IEN) documents describe versions of the Internet Protocol prior to the modern version of IPv4:

- IEN 2 (*Comments on Internet Protocol and TCP*), dated August 1977 describes the need to separate the TCP and Internet Protocol functionalities (which were previously combined.) It proposes the first version of the IP header, using 0 for the version field.

- IEN 26 (*A Proposed New Internet Header Format*), dated February 1978 describes a version of the IP header that uses a 1-bit version field.

- IEN 28 (*Draft Internetwork Protocol Description Version 2*), dated February 1978 describes IPv2.

- IEN 41 (*Internetwork Protocol Specification Version 4*), dated June 1978 describes the first protocol to be called IPv4. The IP header is different from the modern IPv4 header.

- IEN 44 (*Latest Header Formats*), dated June 1978 describes another version of IPv4, also with a header different from the modern IPv4 header.

- IEN 54 (*Internetwork Protocol Specification Version 4*), dated September 1978 is the first description of IPv4 using the header that would be standardized in RFC 760.

The dominant internetworking protocol in the Internet Layer in use today is IPv4; the number 4 is the protocol version number carried in every IP datagram. IPv4 is described in RFC 791 (1981).

Version 5 was used by the Internet Stream Protocol, an experimental streaming protocol.

The successor to IPv4 is IPv6. Its most prominent difference from version 4 is the size of the addresses. While IPv4 uses 32 bits for addressing, yielding c. 4.3 billion (4.3×109) addresses, IPv6 uses 128-bit addresses providing ca. 340 undecillion, or 3.4×1038 addresses. Although adoption of IPv6 has been slow, as of June 2008, all United States government systems have demonstrated basic infrastructure support for IPv6. IPv6 was a result of several years of experimentation and dialog during which various protocol models were proposed, such as TP/IX (RFC 1475), PIP (RFC 1621) and TUBA (TCP and UDP with Bigger Addresses, RFC 1347).

The assignment of the new protocol as IPv6 was uncertain until due diligence revealed that IPv6 had not yet been used previously. Other protocol proposals named *IPv9* and *IPv8* briefly surfaced, but had no affiliation with any international standards body, and have had no support.

On April 1, 1994, the IETF published an April Fool's Day joke about IPv9.

Reliability

The design of the Internet protocols is based on the end-to-end principle. The network infrastructure is considered inherently unreliable at any single network element or transmission medium and is dynamic in terms of availability of links and nodes. No central monitoring or performance measurement facility exists that tracks or maintains the state of the network. For the benefit of reducing network complexity, the intelligence in the network is purposely located in the end nodes.

As a consequence of this design, the Internet Protocol only provides best-effort delivery and its service is characterized as unreliable. In network architectural language, it is a connectionless protocol, in contrast to connection-oriented communications. Various error conditions may occur, such as data corruption, packet loss, duplication and out-of-order delivery. Because routing is dynamic, meaning every packet is treated independently, and because the network maintains no state based on the path of prior packets, different packets may be routed to the same destination via different paths, resulting in out-of-order sequencing at the receiver.

IPv4 provides safeguards to ensure that the IP packet header is error-free. A routing node calculates a checksum for a packet. If the checksum is bad, the routing node discards the packet. Although the Internet Control Message Protocol (ICMP) allows such notification, the routing node is not required to notify either end node of these errors. By contrast, in order to increase performance, and since current link layer technology is assumed to provide sufficient error detection, the IPv6 header has no checksum to protect it.

All error conditions in the network must be detected and compensated by the end nodes of a transmission. The upper layer protocols of the internet protocol suite are responsible for resolving reliability issues. For example, a host may buffer network data to ensure correct ordering before the data is delivered to an application.

Link Capacity and Capability

The dynamic nature of the Internet and the diversity of its components provide no guarantee that any particular path is actually capable of, or suitable for, performing the data transmission requested, even if the path is available and reliable. One of the technical constraints is the size of data packets allowed on a given link. An application must assure that it uses proper transmission characteristics. Some of this responsibility lies also in the upper layer protocols. Facilities exist to examine the maximum transmission unit (MTU) size of the local link and Path MTU Discovery can be used for the entire projected path to the destination. The IPv4 internetworking layer has the capability to automatically fragment the original datagram into smaller units for transmission. In this case, IP provides re-ordering of fragments delivered out of order.

The Transmission Control Protocol (TCP) is an example of a protocol that adjusts its segment size to be smaller than the MTU. The User Datagram Protocol (UDP) and the Internet Control Message Protocol (ICMP) disregard MTU size, thereby forcing IP to fragment oversized datagrams.

An IPv6 network does not perform fragmentation or reassembly, and as per the end-to-end principle, requires end stations and higher-layer protocols to avoid exceeding the network's MTU.

Security

During the design phase of the ARPANET and the early Internet, the security aspects and needs of a public, international network could not be adequately anticipated. Consequently, many Internet protocols exhibited vulnerabilities highlighted by network attacks and later security assessments. In 2008, a thorough security assessment and proposed mitigation of problems was published. The Internet Engineering Task Force (IETF) has been pursuing further studies.

Mobile VoIP

Mobile VoIP or simply mVoIP is an extension of mobility to a Voice over IP network. Two types of communication are generally supported: cordless/DECT/PCS protocols for short range or campus communications where all base stations are linked into the same LAN, and wider area communications using 3G/4G protocols.

There are several methodologies that allow a mobile handset to be integrated into a VoIP network. One implementation turns the mobile device into a standard SIP client, which then uses a data network to send and receive SIP messaging, and to send and receive RTP for the voice path. This methodology of turning a mobile handset into a standard SIP client requires that the mobile handset support, at minimum, high speed IP communications. In this application, standard VoIP protocols (typically SIP) are used over any broadband IP-capable wireless network connection such as EVDO rev A (which is symmetrical high speed — both high speed up and down), HSPA, Wi-Fi or WiMAX.

Another implementation of mobile integration uses a soft-switch like gateway to bridge SIP and RTP into the mobile network's SS7 infrastructure. In this implementation, the mobile handset continues to operate as it always has (as a GSM or CDMA based device), but now it can be controlled by a SIP application server which can now provide advanced SIP-based services to it. Several vendors offer this kind of capability today.

Mobile VoIP will require a compromise between economy and mobility. For example, voice over Wi-Fi offers potentially free service but is only available within the coverage area of a single Wi-Fi access point. Cordless protocols offer excellent voice support and even support base station handoff, but require all base stations to communicate on one LAN as the handoff protocol is generally not supported by carriers or most devices.

High speed services from mobile operators using EVDO rev A or HSPA may have better audio quality and capabilities for metropolitan-wide coverage including fast handoffs among mobile base stations, yet may cost more than Wi-Fi-based VoIP services.

As device manufacturers exploited more powerful processors and less costly memory, smartphones became capable of sending and receiving email, browsing the web (albeit at low rates) and allowing a user to watch TV. Mobile VoIP users were predicted to exceed 100 million by 2012 and InStat projects 288 million subscribers by 2013.

The mobile operator industry business model conflicts with the expectations of Internet users that access is free and fast without extra charges for visiting specific sites, however far away they may be hosted. Because of this, most innovations in mobile VoIP will likely come from campus and corporate networks, open source projects like Asterisk, and applications where the benefits are high enough to justify expensive experiments (medical, military, etc.).

Technologies

Mobile VoIP, like all VoIP, relies on SIP — the standard used by most VoIP services, and now being implemented on mobile handsets and smartphones and an increasing number of cordless phones.

UMA — the Unlicensed Mobile Access Generic Access Network allows VoIP to run over the GSM cellular backbone.

When moving between IP-based networks, as is typically the case for outdoor applications, two other protocols are required:

- IEEE 802.21 handoff, permitting one network to do call setup and initial traffic, handing off to another when the first is about to fall out of range - the underlying network need not be IP-based, but typically the IP stream is guaranteed a certain Quality of Service (QoS) during the handoff process

- IEEE 802.11u call initiation when the initial contact with a network is not one that the user has subscribed to or been in contact with before.

For indoor or campus (cordless phone equivalent) use, the IEEE P1905 protocol establishes QoS guarantees for home area networks: Wi-Fi, Bluetooth, 3G, 4G and wired backbones using AC powerline networking/HomePlug/IEEE P1901, Ethernet and Power over Ethernet/IEEE 802.3af/IEEE 802.3at, MoCA and G.hn. In combination with IEEE 802.21, P1905 permits a call to be initiated on a wired phone and transferred to a wireless one and then resumed on a wired one, perhaps with additional capabilities such as videoconferencing in another room. In this case the use of mobile VoIP enables a continuous conversation that originates, and ends with, a wired terminal device.

An older technology, PCS base station handoff, specifies equivalent capabilities for cordless phones based on 800, 900, 2.4, 5.8 and DECT. While these capabilities were not widely implemented, they did provide the functional specification for handoff for modern IP-based telephony. A phone can in theory offer both PCS cordless and mobile VoIP and permit calls to be handed off from traditional cordless to cell and back to cordless if both the PCS and UMA/SIP/IEEE standards suites are implemented. Some specialized long distance cordless vendors like Senao attempted this but it has not generally caught on. A more popular approach has been full-spectrum handsets that can communicate with any wireless network including mobile VoIP, DECT and satellite phone networks, but which have limited handoff capabilities between networks. The intent of IEEE 802.21 and IEEE 802.11u is that they be added to such phones running iPhone, QNX, Android or other smartphone operating systems, yielding a phone that is capable of communicating with literally any digital network and maintaining a continuous call at high reliability at a low access cost.

Most VoIP vendors implement proprietary technologies that permit such handoff between equipment of their own manufacture, e.g. the Viera system from Panasonic. Typically providing mobility costs more, e.g., the Panasonic VoIP cordless phone system (KX-TGP) costs approximately three times more than its popular DECT PSTN equivalent (KX-TGA). Some companies, including Cisco, offer adapters for analog/DECT phones as alternatives to their expensive cordless.

Industry History

2005

Early experiments proved that VoIP was practical and could be routed by Asterisk even on low-end routers like the Linksys WRT54G series. Suggesting a mesh network (e.g. WDS) composed of such cheap devices could similarly support roaming mobile VoIP phones. These experiments, and others for IP roaming such as Sputnik, were the beginning of the 5G protocol suite including IEEE 802.21 and IEEE 802.11u. At this time, some mobile operators attempted to restrict IP tethering and VoIP use on their networks, often by deliberately introducing high latency into data communications making it useless for voice traffic.

2006

In the summer of 2006, a SIP (Session Initiation Protocol) stack was introduced and a VoIP client in Nokia E-series dual-mode Wi-Fi handsets (Nokia E60, Nokia E61, Nokia E70). The SIP stack and client have since been introduced in many more E and N-series dual-mode Wi-Fi handsets, most notably the Nokia N95 which has been very popular in Europe. Various services use these handsets.

2008

In spring 2008 Nokia introduced a built in SIP VoIP client for the very first time to the mass market device (Nokia 6300i) running Series 40 operating system. Later that year (Nokia 6260 Slide was introduced introducing slightly updated SIP VoIP client. Nokia maintains a list of all phones that have an integrated VoIP client in Forum Nokia.

Aircell's battle with some companies allowing VoIP calls on flights is another example of the growing conflict of interest between incumbent operators and new VoIP operators.

2009

By January 2009 OpenWRT was capable of supporting mobile VoIP applications via Asterisk running on a USB stick. As OpenWRT runs on most Wi-Fi routers, this radically expanded the potential reach of mobile VoIP applications. Users reported acceptable results using G.729 codecs and connections to a "main NAT/Firewall router with a NAT=yes and canreinvite=no.. As such, my asterisk will stay in the audio path and can't redirect the RTP media stream (audio) to go directly from the caller to the callee." Minor problems were also reported: "Whenever there is an I/O activities ... i.e. reading the Flash space (mtdblockd process), this will create some hick-ups (or temporarily losing audio signals)." The combination of OpenWRT and Asterisk is intended as an open source replacement for proprietary PBXes.

The company xG Technology, Inc. had a mobile VoIP and data system operating in the license-free ISM 900 MHz band (902 MHz – 928 MHz). xMax is an end-to-end Internet Protocol (IP) system infrastructure that is currently deployed in Fort Lauderdale, Florida.

2010

In January 2010 Apple Inc. updated the iPhone developer SDK to allow VoIP over cellular networks. iCall became the first App Store app to enable VoIP on the iPhone and iPod Touch over cellular 3G networks.

In second half of 2010 Nokia introduced three new dualmode WiFi capable Series40 handsets (Nokia X3-02, Nokia C3-01 and, Nokia C3-01 Gold Edition) with integrated SIP VoIP that supports HD voice (AMR-WB).

2011

The mainstreaming of VoIP in the small business market led to the introduction of more devices extending VoIP to business cordless users.

Panasonic introduced the KX-TGP base station supporting up to 6 cordless handsets , essentially a VoIP complement to its popular KX-TGA analog phones which likewise support up to 4 cordless handsets. However, unlike the analog system which supports only four handsets in one "conference" on one line, the TGP supports 3 simultaneous network conversations and up to 8 SIP registrations (e.g. up to 8 DID lines or extensions), as well as an Ethernet pass-through port to hook up computers on the same drop. In its publicity Panasonic specifically mentions Digium (founded by the creator of Asterisk), its product Switchvox and Asterisk itself.

Several router manufacturers including TRENDnet and Netgear released sub-$300 Power over Ethernet switches aimed at the VoIP market. Unlike industry standard switches that provided the full 30 watts of power per port, these allowed under 50 watts of power to all four PoE ports combined. This made them entirely suitable for VoIP and other low-power use (Motorola Canopy or security camera or Wi-Fi APs) typical of a SOHO application, or supporting an 8-line PBX, especially in combination with a multi-line handset such as the Panasonic KX-TGP (which does not require a powered port).

Accordingly, by the end of 2011, for under US$3000 it was possible to build an office VoIP system based entirely on cordless technology capable of several hundred meters reach and on Power over Ethernet dedicated wired phones, with up to 8 DID lines and 3 simultaneous conversations per base station, with 24 handsets each capable of communicating on any subset of the 8 lines, plus an unlimited number of softphones running on computers and laptops and smartphones. This compared favourably to proprietary PBX technology especially as VoIP cordless was far cheaper than PBX cordless.

Cisco also released the SPA112, an Analog Telephone Adapter (ATA) to connect one or two standard RJ-11 telephones to an Ethernet, in November 2011, retailing for under US$50. This was a competitive response to major cordless vendors such as Panasonic moving into the business VoIP cordless market Cisco had long dominated, as it suppressed the market for the cordless makers' native VoIP phones and permitted Cisco to argue the business case to spend more on switches and less on terminal devices. However, this solution would not permit the analog phones to access every line of a multi-line PBX, only one hardwired line per phone.

As of late 2011, most cellular data networks were still extremely high latency and effectively useless for VoIP. IP-only providers such as Voipstream had begun to serve urban areas, and alternative approaches such as OpenBTS (open source GSM) were competing with mobile VoIP.

In November 2011, Nokia introduced Nokia Asha 303 with integrated SIP VoIP client that can operate both over WiFi and 3G networks.

2012

In February 2012, Nokia introduced Nokia Asha 302 and in June Nokia Asha 311 both with integrated SIP VoIP client that can operate both over WiFi and 3G networks.

2014

By September 2014, mobile-enabled VoIP (VoLTE) had been launched by T-Mobile US across its national network and by AT&T Mobility in a few markets. Verizon plans to launch its VoLTE service "in the coming weeks," according to media reports in August, 2014. It provides HD Voice, which increases mobile voice quality, and permits optional use of video calling and front and rear-facing cameras. In the future, Verizon's VoLTE is expected to also permit video sharing, chat functionality, and file transfers.

Session Initiation Protocol

The Session Initiation Protocol (SIP) is a communications protocol for signaling and controlling multimedia communication sessions. The most common applications of SIP are in Internet telephony for voice and video calls, as well as instant messaging, over Internet Protocol (IP) networks.

The protocol defines the messages that are sent between endpoints, which govern establishment, termination and other essential elements of a call. SIP can be used for creating, modifying and terminating sessions consisting of one or several media streams. SIP is an application layer protocol designed to be independent of the underlying transport layer. It is a text-based protocol, incorporating many elements of the Hypertext Transfer Protocol (HTTP) and the Simple Mail Transfer Protocol (SMTP).

SIP works in conjunction with several other application layer protocols that identify and carry the session media. Media identification and negotiation is achieved with the Session Description Protocol (SDP). For the transmission of media streams (voice, video) SIP typically employs the Real-time Transport Protocol (RTP) or Secure Real-time Transport Protocol (SRTP). For secure transmissions of SIP messages, the protocol may be encrypted with Transport Layer Security (TLS).

History

SIP was originally designed by Mark Handley, Henning Schulzrinne, Eve Schooler and Jonathan Rosenberg in 1996. The protocol was standardized as RFC 2543 in 1999. In November 2000, SIP was accepted as a 3GPP signaling protocol and permanent element of the IP Multimedia Subsystem (IMS) architecture for IP-based streaming multimedia services in cellular networks. In June 2002 the specification was revised in RFC 3261 and various extensions and clarifications have been published since.

The protocol was designed with the vision to support new multimedia applications. It has been extended for video conferencing, streaming multimedia distribution, instant messaging, presence information, file transfer, fax over IP and online games.

SIP is distinguished by its proponents for having roots in the Internet community rather than in the telecommunications industry. SIP has been standardized primarily by the IETF, while other protocols, such as H.323, have traditionally been associated with the International Telecommunication Union (ITU).

Protocol Operation

SIP is independent from the underlying transport protocol. It runs on the Transmission Control Protocol (TCP), the User Datagram Protocol (UDP) or the Stream Control

Transmission Protocol (SCTP). SIP can be used for two-party (unicast) or multiparty (multicast) sessions.

SIP employs design elements similar to the HTTP request/response transaction model. Each transaction consists of a client request that invokes a particular method or function on the server and at least one response. SIP reuses most of the header fields, encoding rules and status codes of HTTP, providing a readable text-based format.

Each resource of a SIP network, such as a user agent or a voicemail box, is identified by a uniform resource identifier (URI), based on the general standard syntax also used in Web services and e-mail. The URI scheme used for SIP is *sip* and a typical SIP URI has the form *sip:username@domainname* or *sip:username@hostport*, where *domainname* requires DNS SRV records to locate the servers for SIP domain while *hostport* can be an IP address or a fully qualified domain name of the host and port.

If secure transmission is required, the scheme *sips* is used and mandates that each hop over which the request is forwarded up to the target domain must be secured with Transport Layer Security (TLS). The last hop from the proxy of the target domain to the user agent has to be secured according to local policies. TLS protects against attackers who try to listen on the signaling link but it does not provide real end-to-end security to prevent espionage and law enforcement interception, as the encryption is only hop-by-hop and every single intermediate proxy has to be trusted.

SIP works in concert with several other protocols and is only involved in the signaling portion of a communication session. SIP clients typically use TCP or UDP on port numbers 5060 or 5061 to connect to SIP servers and other SIP endpoints. Port 5060 is commonly used for non-encrypted signaling traffic whereas port 5061 is typically used for traffic encrypted with Transport Layer Security (TLS). SIP is primarily used in setting up and tearing down voice or video calls. It also allows modification of existing calls. The modification can involve changing addresses or ports, inviting more participants, and adding or deleting media streams. SIP has also found applications in messaging applications, such as instant messaging, and event subscription and notification. A suite of SIP-related Internet Engineering Task Force (IETF) rules define behavior for such applications. The voice and video stream communications in SIP applications are carried over another application protocol, the Real-time Transport Protocol (RTP). Parameters (port numbers, protocols, codecs) for these media streams are defined and negotiated using the Session Description Protocol (SDP), which is transported in the SIP packet body.

A motivating goal for SIP was to provide a signaling and call setup protocol for IP-based communications that can support a superset of the call processing functions and features present in the public switched telephone network (PSTN). SIP by itself does not define these features; rather, its focus is call-setup and signaling. The features that permit familiar telephone-like operations (i.e. dialing a number, causing a phone to ring, hearing ringback tones or a busy signal) are performed by proxy servers and user agents. Implementation and terminology are different in the SIP world but to the end-user, the behavior is similar.

SIP-enabled telephony networks often implement many of the call processing features of Signaling System 7 (SS7), although the two protocols themselves are very different. SS7 is a centralized protocol, characterized by a complex central network architecture and dumb endpoints (traditional telephone handsets). SIP is a client-server protocol, however most SIP-enabled devices may perform both the client and the server role. In general, session initiator is a client, and the call recipient performs the server function. SIP features are implemented in the communicating endpoints, contrary to traditional SS7 architecture, in which features are implemented in the network core.

Because SIP devices must perform both client and server roles, network communication can be difficult with modern network topologies. When using a connection-oriented protocol like TCP, SIP nominally expects that separate connections will be opened for requests from A to B and requests from B to A. The use of firewalls and NATs interferes with this, as it may not be possible for B to initiate a connection to A, if A is behind a firewall or NAT. SIP allows the original connection from A to B to be used for requests from B to A, but the requests must correctly distinguish between A's private and public addresses and ports; this is also true of requests on connectionless protocols like UDP. To accomplish this, SIP uses extensions like received and rport, and can be paired with other protocols for discovering network topology such as TURN, STUN, and ICE.

Network Elements

SIP defines user agents as well as several types of server network elements. Two SIP endpoints can communicate without any intervening SIP infrastructure. However, this approach is often impractical for public services, which need directory services to locate available nodes in the network.

User Agent

A SIP user agent (UA) is a logical network end-point used to create or receive SIP messages and thereby manage a SIP session. A SIP UA can perform the role of a user agent client (UAC), which sends SIP requests, and the user agent server (UAS), which receives the requests and returns a SIP response. Unlike other network protocols where the roles of client and server are fixed (e.g., a web browser only acts as an HTTP client, and never acts as an HTTP server), in SIP requests can go in either direction, so in almost all cases, a SIP UA must be capable of performing both roles. (If a SIP UA could only perform one role, it could only receive calls and have calls hung up by the peer, but not make calls or hang them up itself, or vice versa.) These roles of UAC and UAS only last for the duration of a SIP transaction.

A SIP phone is an IP phone that implements client and server functions of a SIP user agent and provides the traditional call functions of a telephone, such as dial, answer, reject, call hold, and call transfer. SIP phones may be implemented as a hardware device or as a softphone. As vendors increasingly implement SIP as a standard telephony platform, the distinction between hardware-based and software-based SIP phones is blurred and SIP elements are implemented in the basic firmware functions of many IP-capable devices. Examples are devices from Nokia and BlackBerry.

In SIP, as in HTTP, the user agent may identify itself using a message header field *User-Agent*, containing a text description of the software, hardware, or the product name. The user agent field is sent in request messages, which means that the receiving SIP server can see this information. SIP network elements sometimes store this information, and it can be useful in diagnosing SIP compatibility problems.

Proxy Server

The proxy server is an intermediary entity that acts as both a server and a client for the purpose of making requests on behalf of other clients. A proxy server primarily plays the role of routing, meaning that its job is to ensure that a request is sent to another entity closer to the targeted user. Proxies are also useful for enforcing policy, such as for determining whether a user is allowed to make a call. A proxy interprets, and, if necessary, rewrites specific parts of a request message before forwarding it.

Registrar

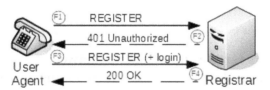

SIP user agent registration to SIP registrar with authentication.

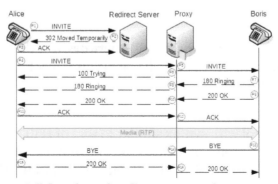

Call flow through redirect server and proxy.

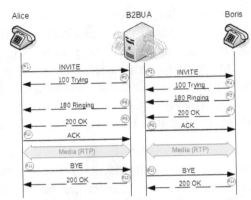

Establishment of a session through a back-to-back user agent.

A registrar is a SIP endpoint that accepts REGISTER requests, recording the address and other parameters from the user agent, and that provides a location service for subsequent requests. The location service links one or more IP addresses to the SIP URI of the registering agent. Multiple user agents may register for the same URI, with the result that all registered user agents receive the calls to the URI.

SIP registrars are logical elements, and are commonly co-located with SIP proxies. To improve network scalability, location services may instead be located with a redirect server.

Redirect Server

A redirect server is a user agent server that generates 3xx (redirection) responses to requests it receives, directing the client to contact an alternate set of URIs. A redirect server allows proxy servers to direct SIP session invitations to external domains.

Session Border Controller

Session border controllers serve as middle boxes between UA and SIP servers for various types of functions, including network topology hiding and assistance in NAT traversal.

Gateway

Gateways can be used to interconnect a SIP network to other networks, such as the public switched telephone network, which use different protocols or technologies.

SIP Messages

SIP is a text-based protocol with syntax similar to that of HTTP. There are two different types of SIP messages: requests and responses. The first line of a request has a *method*, defining the nature of the request, and a Request-URI, indicating where the request should be sent. The first line of a response has a *response code*.

Requests

Requests initiate a SIP transaction between two SIP entities for establishing, controlling, and terminating sessions. Critical methods include the following.

- INVITE: Used to establish a dialog with media exchange between user agents.

- BYE: Terminates an existing session.

- REGISTER: The method implements a location service for user agents, which indicate their address information to the server.

Responses

Responses are send by the user agent server indicating the result of a received request. Several classes of responses are recognized, determined by the numerical range of result codes:

- 1xx: Provisional responses to requests indicate the request was valid and is being processed.

- 2xx: 200-level responses indicate a successful completion of the request. As a response to an INVITE, it indicates a call is established.

- 3xx: This group indicates a redirection is needed for completion of the request. The request has to be completed with a new destination.

- 4xx: The request contained bad syntax or cannot be fulfilled at the server.

- 5xx: The server failed to fulfill an apparently valid request.

- 6xx: This is a global failure, as the request cannot be fulfilled at any server.

Transactions

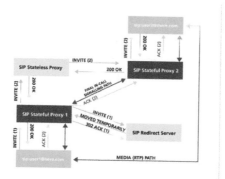

Example: User1's UAC uses an *Invite Client Transaction* to send the initial INVITE (1) message. If no response is received after a timer controlled wait period the UAC may chose to terminate the transaction or retransmit the INVITE. Once a response is received, User1 is confident the INVITE was delivered reliably. User1's UAC must then acknowledge the response. On delivery of the ACK (2) both sides of the transaction are complete. In this case, a dialog may have been established.

SIP defines a transaction mechanism to control the exchanges between participants and deliver messages reliably. A transaction is a state of a session, which is controlled by various timers. Client transactions send requests and server transactions respond to those requests with one or more responses. The responses may include provisional responses, which a response code in the form 1xx, and one or multiple final responses (2xx – 6xx).

Transactions are further categorized as either type *Invite* or type *Non-Invite*. Invite transactions differ in that they can establish a long-running conversation, referred to as a *dialog* in SIP, and so include an acknowledgment (ACK) of any non-failing final response, e.g., *200 OK*.

Because of these transactional mechanisms, unreliable transport protocols, such as the User Datagram Protocol (UDP), are sufficient for SIP operation.

Instant Messaging and Presence

The Session Initiation Protocol for Instant Messaging and Presence Leveraging Extensions (SIMPLE) is the SIP-based suite of standards for instant messaging and presence information. MSRP (Message Session Relay Protocol) allows instant message sessions and file transfer.

Conformance Testing

TTCN-3 test specification language is used for the purposes of specifying conformance tests for SIP implementations. SIP test suite is developed by a Specialist Task Force at ETSI (STF 196). The SIP developer community meets regularly at the SIP Forum SIPit events to test interoperability and test implementations of new RFCs.

Performance Testing

When developing SIP software or deploying a new SIP infrastructure, it is very important to test capability of servers and IP networks to handle certain call load: number of concurrent calls and number of calls per second. SIP performance tester software is used to simulate SIP and RTP traffic to see if the server and IP network are stable under the call load. The software measures performance indicators like answer delay, answer/seizure ratio, RTP jitter and packet loss, round-trip delay time.

Applications

A *SIP connection* is a marketing term for voice over Internet Protocol (VoIP) services offered by many Internet telephony service providers (ITSPs). The service provides routing of telephone calls from a client's private branch exchange (PBX) telephone system to the public switched telephone network (PSTN). Such services may simplify corporate information system infrastructure by sharing Internet access for voice and data, and removing the cost for Basic Rate Interface (BRI) or Primary Rate Interface (PRI) telephone circuits.

Many VoIP phone companies allow customers to use their own SIP devices, such as SIP-capable telephone sets, or softphones.

SIP-enabled video surveillance cameras can make calls to alert the owner or operator that an event has occurred; for example, to notify that motion has been detected out-of-hours in a protected area.

SIP is used in audio over IP for broadcasting applications where it provides an interoperable means for audio interfaces from different manufacturers to make connections with one another.

Implementations

The U.S. National Institute of Standards and Technology (NIST), Advanced Networking Technologies Division provides a public-domain Java implementation that serves as a reference implementation for the standard. The implementation can work in proxy server or user agent scenarios and has been used in numerous commercial and research projects. It supports RFC 3261 in full and a number of extension RFCs including RFC 6665 (event notification) and RFC 3262 (reliable provisional responses).

Numerous other commercial and open-source SIP implementations exist. List of SIP software.

SIP-ISUP Interworking

SIP-I, or the Session Initiation Protocol with encapsulated ISUP, is a protocol used to create, modify, and terminate communication sessions based on ISUP using SIP and IP networks. Services using SIP-I include voice, video telephony, fax and data. SIP-I and SIP-T are two protocols with similar features, notably to allow ISUP messages to be transported over SIP networks. This preserves all of the detail available in the ISUP header, which is important as there are many country-specific variants of ISUP that have been implemented over the last 30 years, and it is not always possible to express all of the same detail using a native SIP message. SIP-I was defined by the ITU-T, whereas SIP-T was defined via the IETF RFC route.

Encryption

The increasing concerns about the security of calls that run over the public Internet has made SIP encryption more popular and, in fact more desired. Because VPN is not an option for most service providers, most service providers that offer secure SIP (SIPS) connections use TLS for securing signaling. The relationship between SIP (port 5060) and SIPS (port 5061), is similar to that as for HTTP and HTTPS, and uses URIs in the form "sips:user@example.com". The media streams, which occur on different connections to the signaling stream, can be encrypted with SRTP. The key exchange for SRTP is performed with SDES (RFC 4568), or the newer and often more user friendly ZRTP (RFC 6189), which can automatically upgrade RTP to SRTP using dynamic key exchange (and a verification phrase). One can also add a MIKEY (RFC 3830) exchange to SIP and in that way determine session keys for use with SRTP.

IEEE 802.21

802.21 is an IEEE standard published in 2008. The standard supports algorithms enabling seamless handover between networks of the same type as well as handover between different network types also called Media independent handover (MIH) or vertical handover. The standard provides information to allow handing over to and from 802.3, 802.11, 802.15, 802.16, 3GPP and 3GPP2 networks through different handover mechanisms.

The IEEE 802.21 working group started work in March 2004. More than 30 companies have joined the working group. The group produced a first draft of the standard including the protocol definition in May 2005. The standard was published January 2009.

Reasons for 802.21

Cellular networks and 802.11 networks employ handover mechanisms for handover within the same network type (aka horizontal handover). Mobile IP provides handover mechanisms for handover across subnets of different types of networks, but can be slow in the process. Current 802 standards do not support handover between different types of networks. They also do not provide triggers or other services to accelerate mobile IP based handovers. Moreover, existing 802 standards provide mechanisms for detecting and selecting network access points, but do not allow for detection and selection of network access points in a way that is independent of the network type.

Some of the Expectations

- Allow roaming between 802.11 networks and 3G cellular networks.

- Allow users to engage in ad hoc teleconferencing.

- Apply to both wired and wireless networks, likely the same list as IEEE P1905 specifies to cooperate in software-defined networking

- Allow for use by multiple vendors and users.

- o Compatibility and conformance with other IEEE 802 standards especially 802.11u unknown user authentication and 802.11s ad hoc wireless mesh networking.
- Include definitions for managed objects that are compatible with management standards like SNMP.
- Although security algorithms and security protocols will not be defined in the standard, authentication, authorization, and network detection and selection will be supported by the protocol.

Implementation

Many vendors are building wireless products and participating in the development of the 802.21 standard. Current technologies such as 802.11 that accomplish handover use software to accomplish handovers and suggest that software will also be the way that handover will be implemented by 802.21. The use of software as a means to implement 802.21 should not cause large increases in the cost of networking devices. An open source software implementation is provided by ODTONE.

Examples

- A user should be able to unplug from an 802.3 network and get handed off to an 802.11 network.
- A cellular phone user in the midst of a call should be able to enter an 802.11 network hotspot and be seamlessly handed off from a GSM network to the 802.11 network and back again when leaving the hotspot.

Similar Technologies

Unlicensed Mobile Access (UMA) technology is basically a mobile-centric version of 802.21. UMA is said to provide roaming and handover between GSM, UMTS, Bluetooth and 802.11 networks. Since June 19, 2005, UMA is a part of the ETSI 3GPP standardization process under the GAN (Generic Access Network) Group.

The Evolved Packet Core (EPC) architecture for Next Generation Mobile Networks (3GPP Rel.8 and newer) provides the Access Network Discovery and Selection Function element (ANDSF) (3GPP TS 23.402 and 3GPP TS 24.312). Its S14 interface provides the communication path between the Core Network and the User Endpoint device on which to exchange discovery information and inter-system mobility policies, enabling as such a network suggested reselection of access networks.

WiOptiMo technology enables any application running on a device to use the best internet connection among all the wired/wireless access providers available, guaranteeing persistence in case of weak or no signal and managing the switch among them (when needed/convenient) in a transparent way, without interrupting the active application/ session. For more information you can read A Cross-Layering and Autonomic Approach to Optimized Seamless Handover or "WiSwitch: Seamless Handover between Multi-Provider Networks" on IEEE Explore.

IEEE 802.11u

IEEE 802.11u-2011 is an amendment to the IEEE 802.11-2007 standard to add features that improve interworking with external networks.

802.11 is a family of IEEE technical standards for mobile communication devices such as laptop computers or multi-mode phones to join a wireless local area network (WLAN) widely used in the home, public hotspots and commercial establishments.

The IEEE 802.11u standard was published on February 25, 2011.

Some Amendment Added to IEEE 802.11

Network Discovery and Selection

1. Provides for the discovery of suitable networks (preassociation) through the advertisement of access network type {private network, free public network, for-fee public network}, roaming consortium, and venue information.

2. Generic Advertisement Service (GAS), which provides for Layer 2 transport of an advertisement protocol's frames between a mobile device and a server in the network prior to authentication. The access point is responsible for the relay of a mobile device's query to a server in the carrier's network and for delivering the server's response back to the mobile.

3. Provides Access Network Query Protocol (ANQP), which is a query and response protocol used by a mobile device to discover a range of information, including the hotspot operator's domain name (a globally unique, machine searchable data element); roaming partners accessible via the hotspot along with their credential type and EAP method supported for authentication; IP address type availability (for example, IPv4, IPv6); and other metadata useful in a mobile device's network selection process.

QoS Map Distribution

This provides a mapping between the IP's differentiated services code point (DSCP) to over-the-air Layer 2 priority on a per-device basis, facilitating end-to-end QoS.

For Users who are not Pre-authorized

IEEE 802.11 currently makes an assumption that a user's device is pre-authorized to use the network. IEEE 802.11u covers the cases where that device is not pre-authorized. A network will be able to allow access based on the user's relationship with an external network (e.g. hotspot roaming agreements), or indicate that online enrollment is possible, or allow access to a strictly limited set of services such as emergency services (client to authority and authority to client.)

From a user perspective, the aim is to improve the experience of a traveling user who turns on a laptop in a hotel many miles from home, or uses a mobile device to place a phone call. Instead of being presented with a long list of largely meaningless SSIDs the user could be presented with a list of networks, the services they provide, and the conditions under which the user could access them. 802.11u is central to the adoption of UMA and other approaches to network mobile devices.

Encourages Mesh Deployment

Because a relatively sophisticated set of conditions can be presented, arbitrary contracts could be presented to the user, and might include providing information on motive, demographics or geographic origin of the user. As such data is valuable to tourism promotion and other public functions, 802.11u is thought to motivate more extensive deployment of IEEE 802.11s mesh networks.

Mobile Cellular Network Off-load to Wi-Fi

Mobile users, whose devices can move between 3G and Wi-Fi networks at a low level using 802.21 handoff, also need a unified and reliable way to authorize their access to all of those networks. 802.11u provides a common abstraction that all networks regardless of protocol can use to provide a common authentication experience.

Mandatory Requirements

The IEEE 802.11u requirements specification contains requirements in the areas of enrollment, network selection, emergency call support, emergency alert notification, user traffic segmentation, and service advertisement.

Implementation

Hotspot 2.0

The Wi-Fi Alliance uses IEEE 802.11u in its "Wi-Fi Certified Passpoint" program, also known as "Hotspot 2.0". Apple devices running iOS 7 support Hotspot 2.0.

EAP-TLS

There have been proposals to use IEEE 802.11u for access points to signal that they allow EAP-TLS using only server-side authentication. Unlike most TLS implementations of HTTPS, such as major web browsers, the majority of implementations of EAP-TLS require client-side X.509 certificates without giving the option to disable the requirement, even though the standard does not mandate their use, which some have identified as having the potential to dramatically reduce adoption of EAP-TLS and prevent "open" but encrypted access points.

Media Gateway Control Protocol

The Media Gateway Control Protocol (MGCP) is a signaling and call control communications protocol used in voice over IP (VoIP) telecommunication systems. It implements the media gateway control protocol architecture for controlling media gateways on Internet Protocol (IP) networks connected to the public switched telephone network (PSTN). The protocol is a successor to the Simple Gateway Control Protocol (SGCP), which was developed by Bellcore and Cisco, and the Internet Protocol Device Control (IPDC).

The methodology of MGCP reflects the structure of the PSTN with the power of the network residing in a call control center softswitch which is analogous to the central office in the telephone network. The endpoints are low-intelligence devices, mostly executing control commands and providing result indications in response. The protocol represents a decomposition of other VoIP models, such as H.323, in which the H.323 Gatekeeper, have higher levels of signaling intelligence.

MGCP is a text-based protocol consisting of commands and responses. It uses the Session Description Protocol (SDP) for specifying and negotiating the media streams to be transmitted in a call session and the Real-time Transport Protocol (RTP) for framing the media streams.

Architecture

The media gateway control protocol architecture and its methodologies and programming interfaces are described in RFC 2805.

MGCP is a master/slave protocol that allows a call control device such as a Call Agent to take control of a specific port on a media gateway. In MGCP context media gateway controller is referred to as call agent. This has the advantage of centralized gateway administration and provides for largely scalable IP Telephony solutions. The distributed system is composed of a call agent, at least one media gateway (MG) that performs the conversion of media signals between circuits and packets switched networks, and at least one signaling gateway (SG) when connected to the PSTN.

MGCP assumes a call control architecture where there is limited intelligence at the edge (endpoints, media gateways) and intelligence at the core Call Agent. The MGCP assumes that Call Agents, will synchronize with each other to send coherent commands and responses to the gateways under their control.

The Call Agent uses MGCP to tell the media gateway which events should be reported to the Call Agent, how endpoints should be inter-connected, and which signals should be activated on the endpoints.

MGCP also allows the Call Agent to audit the current state of endpoints on a media gateway.

The media gateway uses MGCP to report events, such as off-hook or dialed digits, to the Call Agent.

While any signaling gateway is usually on the same physical switch as a media gateway, there is no such need. The Call Agent does not use MGCP to control the Signaling Gateway; rather, SIGTRAN protocols are used to backhaul signaling between the Signaling Gateway and Call Agent.

Multiple Call Agents

Typically, a media gateway is configured with a list of Call Agents from which it may accept programming (where that list normally comprises only one or two Call Agents).

In principle, event notifications may be sent to different Call Agents for each endpoint on the gateway (as programmed by the Call Agents, by setting the NotifiedEntity parameter). In practice, however, it is usually desirable that at any given moment all endpoints on a gateway should be controlled by the same Call Agent; other Call Agents are available only to provide redundancy in the event that the primary Call Agent fails, or loses contact with the media gateway. In the event of such a failure it is the backup Call Agent's responsibility to reprogram the MG so that the gateway comes under the control of the backup Call Agent. Care is needed in such cases; two Call Agents may know that they have lost contact with one another, but this does not guarantee that they are not both attempting to control the same gateway. The ability to audit the gateway to determine which Call Agent is currently controlling can be used to resolve such conflicts.

MGCP assumes that the multiple Call Agents will maintain knowledge of device state among themselves (presumably with an unspecified protocol) or rebuild it if necessary (in the face of catastrophic failure). Its failover features take into account both planned and unplanned outages.

Protocol Overview

MGCP recognizes three essential elements of communication, the *media gateway controller* (call agent), the media gateway *endpoint*, and *connections* between these entities. A media gateway may host multiple endpoints and each endpoint should be able to engage in multiple connections. Multiple connections on the endpoints support calling features such as call waiting and three-way calling.

MGCP is a text-based protocol using a command and response model. Commands and responses are encoded in messages that are structured and formatted with the whitespace characters space, horizontal tab, carriage return, and linefeed, and the colon and the full stop. Messages are transmitted using the User Datagram Protocol (UDP). Media gateways use the port number 2427, and call agents use 2727 by default.

The message sequence of command (or request) and its response is known as a transaction, which is identified by the numerical Transaction Identifier exchanged in each transaction. The protocol specification defines nine standard commands that are distinguished by a four-letter command verb: AUEP, AUCX, CRCX, DLCX, EPCF, MDCX, NTFY, RQNT, and RSIP. Responses begin with a three-digit numerical response code that identifies the outcome or result of the transaction.

Two verbs are used by a call agent to query the state of an endpoint:

- AUEP: Audit Endpoint

- AUCX: Audit Connection

Three verbs are used by a call agent to manage the connection between a media gateway.

- CRCX: Create Connection

- DLCX: Delete Connection. An endpoint may also terminate a connection with this command.

- MDCX: Modify Connection

One verb is used by a call agent to request notification of events on the endpoint, and to apply signals:

- RQNT: Request for Notification

One verb is used by a call agent to modify coding characteristics expected by the line side of the endpoint:

- EPCF: Endpoint Configuration

One verb is used by an endpoint to indicate to the call agent that it has detected an event for which the call agent had previously requested notification with the RQNT command:

- NTFY: Notify

One verb is used by an endpoint to indicate to the call agent that it is in the process of restarting:

- RSIP: Restart In Progress

Standards Documents

- RFC 3435 - Media Gateway Control Protocol (MGCP) Version 1.0 (this supersedes RFC 2705)

- RFC 3660 - Basic Media Gateway Control Protocol (MGCP) Packages (informational)

- RFC 3661 - Media Gateway Control Protocol (MGCP) Return Code Usage

- RFC 3064 - MGCP CAS Packages

- RFC 3149 - MGCP Business Phone Packages

- RFC 3991 - Media Gateway Control Protocol (MGCP) Redirect and Reset Package

- RFC 3992 - Media Gateway Control Protocol (MGCP) Lockstep State Reporting Mechanism (informational)

- RFC 2805 - Media Gateway Control Protocol Architecture and Requirements

- RFC 2897 - Proposal for an MGCP Advanced Audio Package

Megaco

Another implementation of the media gateway control protocol architecture is the H.248/Megaco protocol, a collaboration of the Internet Engineering Task Force (RFC 3525) and the International Telecommunication Union (Recommendation H.248.1). Both protocols follow the guidelines of the overlying media gateway control protocol architecture, as described in RFC 2805. However, the protocols are incompatible due to differences in protocol syntax and underlying connection model.

Simple Gateway Control Protocol

Simple Gateway Control Protocol (SGCP) is a communications protocol used within a Voice over Internet Protocol (VoIP) system. It has been superseded by MGCP, an implementation of the Media Gateway Control Protocol Architecture.

The Simple Gateway Control Protocol was published in 1998 by Christian Huitema and

Mauricio Arango, as part of the development of the "Call Agent Architecture" at Telcordia Technologies (formerly Bellcore). In this architecture a call-control elements, Media Gateway Controllers (MGCs) or the Call Agent, controls trunking, residential, and access-type VoIP "media gateways" and receives telephony signaling requests through a "signalling gateway".Although these gateways target different market segments, all of them convert time-division multiplexing (TDM) voice to packet voice. Later implementation of the architecture refer to the "Call Agent" as a "Softswitch".

SGCP was intended to be compatible with the Session Initiation Protocol (SIP), enabling the Call Agent to relay calls between a Voice over IP network using SIP and a traditional telephone network. The SGCP commands are encoded with a syntax somewhat comparable to the SIP or HTTP headers. They carry a payload describing the voice over IP media stream. This payload is encoded using the same "session description protocol" (SDP) as SIP.

SGCP was merged with the Internet Protocol Device Control (IPDC) proposal sponsored by Level3 Communications. This led to the definition of the Media Gateway Control Protocol Version 1.0, jointly submitted to the IETF by the authors of SGCP and IPDC in November 1998.

References

- Johnston, Alan B. (2004). SIP: Understanding the Session Initiation Protocol, Second Edition. Artech House. ISBN 1-58053-168-7.

- Azzedine (2006). Handbook of algorithms for wireless networking and mobile computing. CRC Press. p. 774. ISBN 978-1-58488-465-1.

- Porter, Thomas; Andy Zmolek; Jan Kanclirz; Antonio Rosela (2006). Practical VoIP Security. Syngress. pp. 76–77. ISBN 978-1-59749-060-3.

- "WIRELESS: Carriers look to IP for back haul". www.eetimes.com. EE Times. Archived from the original on August 9, 2011. Retrieved 8 April 2015.

- "Mobile's IP challenge". www.totaltele.com. Total Telecom Online. Archived from the original on February 17, 2006. Retrieved 8 April 2015.

- Mike Dano, FierceWireless, "After VoLTE, what is the future of the telephone call?" Mobile Internet Solutions, September 4, 2014

- "Level 3 Communications, Bellcore Announce Merger of Protocol Specifications for Voice Over IP". Level 3 Communications. Retrieved 8 June 2012.

- 111th Congress (2009) (January 7, 2009). "S. 30 (111th)". Legislation. GovTrack.us. Retrieved June 25, 2012. Truth in Caller ID Act of 2009

- Federal Communications Commission (June 22, 2011). "Rules and Regulation Implementing the Truth in Caller ID Act of 2009". fcc.gov. Retrieved June 25, 2012.

Various Device Bit Rates

Bit rates are the indicators of speed and the higher the bit rate the faster the network. It is essential to study the bit rates of devices to understand the most efficient technology available and also as a comparison tool. The content of this chapter is designed to facilitate a better understanding of the bit rates of various devices that are currently in use. The topics discussed in the chapter are of great importance to broaden the existing knowledge on WiMAX technology.

List of Device Bit Rates

This is a list of device bit rates, is a measure of information transfer rates, or digital bandwidth capacity, at which digital interfaces in a computer or network can communicate over various kinds of buses and channels. The distinction can be arbitrary between a *computer bus*, often closer in space, and larger telecommunications networks. Many device interfaces or protocols (e.g., SATA, USB, SAS, PCIe) are used both inside many-device boxes, such as a PC, and one-device-boxes, such as a hard drive enclosure. Accordingly, this page lists both the internal ribbon and external communications cable standards together in one sortable table.

Factors Limiting Actual Performance, Criteria for Real Decisions

Most of the listed rates are theoretical maximum throughput measures; in practice, the actual effective throughput is almost inevitably lower in proportion to the load from other devices (network/bus contention), physical or temporal distances, and other overhead in data link layer protocols etc. The maximum goodput (for example, the file transfer rate) may be even lower due to higher layer protocol overhead and data packet retransmissions caused by line noise or interference such as crosstalk, or lost packets in congested intermediate network nodes. All protocols lose something, and the more robust ones that deal resiliently with very many failure situations tend to lose more maximum throughput to get higher total long term rates.

Device interfaces where one bus transfers data via another will be limited to the throughput of the slowest interface, at best. For instance, SATA 6G controllers on one PCIe 5G channel will be limited to the 5G rate and have to employ more channels to get around this problem. Early implementations of new protocols very often have this kind of problem. The physical phenomena on which the device relies (such as spinning platters in a hard drive) will also impose limits; for instance, no spinning platter shipping in 2009 saturates SATA II (3 Gbit/s), so moving from this 3 Gbit/s interface to USB3 at 4.8 Gbit/s for one spinning drive will result in no increase in realized transfer rate.

Contention in a wireless or noisy spectrum, where the physical medium is entirely out of the control of those who specify the protocol, requires measures that also use up throughput. Wireless devices, BPL, and modems may produce a higher line rate or gross bit rate, due to error-correcting codes and other physical layer overhead. It is extremely common for throughput to be far less than half of theoretical maximum, though the more reycent technologies (notably BPL) employ preemptive spectrum analysis to avoid this and so have much more potential to reach actual gigabit rates in practice than prior modems.

Another factor reducing throughput is deliberate policy decisions made by Internet service providers that are made for contractual, risk management, aggregation saturation, or marketing reasons. Examples are rate limiting, bandwidth throttling, and the assignment of IP addresses to groups. These practices tend to minimize the throughput available to every user, but maximize the number of users that can be supported on one backbone.

Furthermore, chips are often not available in order to implement the fastest rates. AMD, for instance, does not support the 32-bit HyperTransport interface on any CPU it has shipped as of the end of 2009. Additionally, WiMax service providers in the US typically support only up to 4 Mbit/s as of the end of 2009.

Choosing service providers or interfaces based on theoretical maxima is unwise, especially for commercial needs. A good example is large scale data centers, which should be more concerned with price per port to support the interface, wattage and heat considerations, and total cost of the solution. Because some protocols such as SCSI and Ethernet now operate many orders of magnitude faster than when originally deployed, scalability of the interface is one major factor, as it prevents costly shifts to technologies that are not backward compatible. Underscoring this is the fact that these shifts often happen involuntarily or by surprise, especially when a vendor abandons support for a proprietary system.

Conventions

By convention, bus and network data rates are denoted either in bits per second (bit/s) or bytes per second (B/s). In general, parallel interfaces are quoted in B/s and serial in bit/s. The more commonly used is shown below in bold type.

On devices like modems, bytes may be more than 8 bits long because they may be individually padded out with additional start and stop bits; the figures below will reflect this. Where channels use line codes (such as Ethernet, Serial ATA and PCI Express), quoted rates are for the decoded signal.

The figures below are simplex data rates, which may conflict with the duplex rates vendors sometimes use in promotional materials. Where two values are listed, the first value is the downstream rate and the second value is the upstream rate.

All quoted figures are in metric decimal units. Note that these aren't the traditional binary prefixes for memory size. These decimal prefixes have long been established in data communications. This occurred before 1998 when IEC and other organizations introduced new binary prefixes and attempted to standardize their use across all computing applications.

Bandwidths

The figures below are grouped by network or bus type, then sorted within each group from lowest to highest bandwidth; gray shading indicates a lack of known implementations.

Wireless Networks

802.11 networks in infrastructure mode are half-duplex; all stations share the medium. In infrastructure or access point mode, all traffic has to pass through an Access Point (AP). Thus, two stations on the same access point that are communicating with each other must have each and every frame transmitted twice: from the sender to the access point, then from the access point to the receiver. This approximately halves the effective bandwidth.

802.11 networks in ad hoc mode are still half-duplex, but devices communicate directly rather than through an access point. In this mode all devices must be able to "see" each other, instead of only having to be able to "see" the access point.

Dynamic Random-access Memory

The table below shows values for PC memory module types. These modules usually combine multiple chips on one circuit board. SIMM modules connect to the computer via an 8 bit or 32 bit wide interface. DIMM modules connect to the computer via a 64 bit wide interface. Some other computer architectures use different modules with a different bus width.

In a single-channel configuration, only one module at a time can transfer information to the CPU. In multi-channel configurations, multiple modules can transfer information to the CPU at the same time, in parallel. FPM, EDO, SDR, and RDRAM memories were not commonly installed in a dual-channel configuration. DDR and DDR2 memory are usually installed in single or dual-channel configuration. DDR3 memory are installed in single, dual, tri, and quad-channel configurations. Bit rates of multi-channel configurations are the product of the module bit-rate (given below) and the number of channels.

Graphics Processing Units' RAM

RAM memory modules are also utilised by graphics processing units; however, memory modules for those differs somewhat, particularly with lower power requirements, and is specialised to serve GPUs: for example, the introduction of GDDR3, which was fundamentally based on DDR2. Every graphics memory chip is directly connected to the GPU (point-to-point). The total GPU memory bus width varies with the number of memory chips and the number of lanes per chip. For example, GDDR5 specifies either 16 or 32 lanes per "device" (chip), while GDDR5X specifies 64 lanes per chip. Over the years, bus widths rose from 64-bit to 512-bit and beyond - e.g. HBM is 1024 bits wide. Because of this variability, graphics memory speeds are sometimes compared per pin. For direct comparison to the values for 64-bit modules shown above, video RAM is compared here in 64-lane lots, corresponding to two chips for those devices with 32-bit widths. In 2012, high-end GPUs use 8 or even 12 chips with 32 lanes each, for a total memory bus width of 256 or 384 bits. Combined with a transfer rate per pin of 5 GT/s or more, such cards can reach 240 GB/s or more.

RAM frequencies used for a given chip technology vary greatly. Where single values are given below, they are examples from high-end cards. Since many cards have more than one pair of chips, the total bandwidth is correspondingly higher. For example, high-end cards often have eight chips each 32-bits wide, so the total bandwidth for such cards is four times the value given below.

Digital Video Interconnects

Data rates given are from the video source (e.g., video card) to receiving device (e.g., monitor) only. Out of band and reverse signaling channels are not included.

Permissions

Index